The series "Advances in Intelligent Systems and Computing" contains publications on theory, applications, and design methods of Intelligent Systems and Intelligent Computing. Virtually all disciplines such as engineering, natural sciences, computer and information science, ICT, economics, business, e-commerce, environment, healthcare, life science are covered. The list of topics spans all the areas of modern intelligent systems and computing such as: computational intelligence, soft computing including neural networks, fuzzy systems, evolutionary computing and the fusion of these paradigms, social intelligence, ambient intelligence, computational neuroscience, artificial life, virtual worlds and society, cognitive science and systems, Perception and Vision, DNA and immune based systems, self-organizing and adaptive systems, e-Learning and teaching, human-centered and human-centric computing, recommender systems, intelligent control, robotics and mechatronics including human-machine teaming, knowledge-based paradigms, learning paradigms, machine ethics, intelligent data analysis, knowledge management, intelligent agents, intelligent decision making and support, intelligent network security, trust management, interactive entertainment, Web intelligence and multimedia.

The publications within "Advances in Intelligent Systems and Computing" are primarily proceedings of important conferences, symposia and congresses. They cover significant recent developments in the field, both of a foundational and applicable character. An important characteristic feature of the series is the short publication time and world-wide distribution. This permits a rapid and broad dissemination of research results.

More information about this series at http://www.springer.com/series/11156

Jyotsna Kumar Mandal · Devadatta Sinha
J. P. Bandopadhyay
Editors

Contemporary Advances in Innovative and Applicable Information Technology

Proceedings of ICCAIAIT 2018

 Springer

Editors
Jyotsna Kumar Mandal
Department Computer Science
    and Engineering
University of Kalyani
Kalyani, West Bengal, India

J. P. Bandopadhyay
Institute of Radio Physics
    and Electronics
University of Calcutta
Kolkata, West Bengal, India

Devadatta Sinha
Department Computer Science
    and Engineering
University of Calcutta
Kolkata, West Bengal, India

ISSN 2194-5357   ISSN 2194-5365   (electronic)
Advances in Intelligent Systems and Computing
ISBN 978-981-13-1539-8   ISBN 978-981-13-1540-4   (eBook)
https://doi.org/10.1007/978-981-13-1540-4

Library of Congress Control Number: 2018948690

This Springer imprint is published by the registered company Springer Nature Singapore Pte Ltd.
The registered company address is: 152 Beach Road, #21-01/04 Gateway East, Singapore 189721, Singapore

Preface

First International Conference on Contemporary Advances in Innovative and Applicable Information Technology (ICCAIAIT 2018) has been organized during 24–25 March 2018 at the college campus of Kingston Educational Institute, Berunanpukuria, Barasat, Kolkata 700126, West Bengal, India. The conference was organized in collaboration with Computer Society of India (CSI), Division IV (Communication), which was the technical sponsor of the event. The technical co-sponsors were the Institution of Engineering and Technology (IET) and the Institute of Electrical and Electronics Engineers (IEEE), Kolkata Section. The proceedings of the conference is published by Springer, the publication partner of the event in the AISC series of Springer Nature.

There were tutorials, keynote addresses, invited talks and general speeches of interest, talks from industries, paper presentations and panel discussions in multiple/track sessions, covering diverse topics of interest. The proceedings of the selected and presented papers is published in this AISC series of Springer Nature by Springer at free of cost. We are thankful to the authority of Springer for their active association with CSI.

The papers are checked for similarities multiple times through iThenticate, for which access is provided by Springer, followed by updation by the author and double-blind reviews. EasyChair scoring scheme has been adopted to select 23 quality papers based on reviews and EasyChair scoring. We received papers from various premier institutes as well as some papers from the USA and UK. The presented papers were modified as per suggestions from session adjudicators, and tailored papers were uploaded to Springer after checking similarities again. Springer scrutinized all uploaded papers, and finally, they selected the papers for publication. We are grateful to the contributors and reviewers for their effort.

The single volume of this proceeding contains chapters like computational intelligence, data analytics, nature-inspired computing, circuit system and devices, wireless, mobile and cloud computing and social network.

Some papers in this volume are hierarchical image cryptosystem, intelligent Web service searching, computational intelligence-based neural session key generation, computation of peak tunnelling current density in resonant tunnelling diode, organic electricity from Zn/Cu-PKL electrochemical cell, investigation of data set from diabetic retinopathy, probabilistic sink placement strategy in wireless sensor network, social media activity of local traffic police department, potential customer base identification in social media, etc.

We hope this volume will be useful and contributes actively towards the holistic progress of the civilization. This volume will be value-added bits and pieces for academicians, researchers and young promising engineers.

Kalyani, India   Jyotsna Kumar Mandal
                                                    Professor, CSE, University of Kalyani
Kolkata, India   Devadatta Sinha
                                    Former Professor, CSE, University of Calcutta
Kolkata, India   J. P. Bandopadhyay
                                    Emeritus Professor, University of Calcutta

List of Committee Members

Chief Patron

Smt. Uma Bhattacharjee, Secretary, Kingston Educational Institute, India

Patron

Mr. Tipam Bhattacharjee, President, Kingston Educational Institute, India

General Chair

Prof. Dilip Kumar Sinha, Former Vice Chancellor of Visva Bharati, India

Editorial Board

Dr. Jyotsna Kumar Mandal, University of Kalyani (Corresponding Editor), India
Dr. Devadatta Sinha, Former Professor of CSE, University of Calcutta, India
Prof. J. P. Bandopadhyay, Emeritus Professor, University of Calcutta; Academic
  Chairman, Kingston Educational Institute, India

Organizing Chairs

Prof. J. P. Bandopadhyay, Emeritus Professor, University of Calcutta; Academic
Chairman, Kingston Educational Institute, India
Mr. Devaprasanna Sinha, RVP—Region-II, CSI, India
Prof. Asish Mukhopadhyay, Ex-Group Director, SRGI, Jhansi; Advisor, Kingston
Polytechnic College

Organizing Co-chairs

Dr. Manishankar Chakraborty, Advisor, Kingston Educational Institute, India
Prof. Diptarup Bandyopadhyay, Principal, Kingston Polytechnic College, India
Mr. S. C. Rudra, Former Director, All India Radio, Akashvani Bhavan, Kolkata

International Advisory Committee

Dr. Takaaki Goto, Ryutsu Keizai University, Ryugasaki, Japan
Dr. Amlan Chatterjee, California State University, USA

Dr. Sanjay Mahapatra, National President of CSI, India
Dr. Goutam Mahapatra, Vice President of CSI, India
Dr. A. K. Nayak, Secretary of CSI, India
Dr. Nabendu Chaki, Calcutta University, India
Dr. Amit Banerjee, Shizuoka University, Japan
Dr. Soumya Sen, Calcutta University, India
Dr. Durgesh Misra, Chairman, Div IV of CSI, India
Dr. Vipin Tyagi, RVP—Region-III, CSI, India
Dr. Paramartha Dutta, Visva Bharati, India
Dr Amlan Chakrabarti, Calcutta University, India
Dr. Sankhayan Choudhury, Calcutta University, India
Dr. Achintya Das, Kalyani Government Engineering College, India
Dr. Siddheswar Maikap, Chang Gung University, Taiwan, R.O.C
Dr. Qing Hao, The University of Arizona, USA
Dr. Gora Chand Dutta, Michigan State University, USA
Dr. Somnath Chattopadhyay, California State University, Northridge, USA
Dr. Pierre Mialhe, University of Perpignan, France
Dr. Gautam Kumar Dalapati, IMRE, A*Star, Singapore
Prof. Hiroshi Inokawa, Shizuoka University, Japan
Dr. Al-Sakib Khan Pathan, School of IT, Geelong Waurn, Malaysia
Dr. Toni Janevski, Ss. Cyril and Methodius University, Macedonia
Dr. Sunil Kumar Jha, University of Information Science and Technology, Macedonia
Dr. Rudra Dutta, North Carolina State University, USA
Dr. P. T. Kulkarni, Pune Institute of Computer Technology, India

Conference Chairs

Dr. Anirban Basu, President of CSI, India
Mr. Sanjoy Mahapatra, Vice President of CSI, India

Organizing Secretaries

Mr. Partha Ghosh, Asst. Prof., Kingston School of Management & Science, India
Mr. Rajib Kr. Sanyal, Asst. Prof., Kingston School of Management & Science, India
Mr. Naba Kr. Bera, Administrative Officer, Kingston Law College, India

Joint Organizing Secretaries

Mr. Sovonesh Pal, Dean, Faculty of Engineering and Technology, Kingston Polytechnic College, India
Smt. Priyanka Bhattacharya, Asst. Prof., Kingston School of Management & Science, India

Technical Programme Committee

Technical Programme Committee Chairs

Dr. Jyotsna Kumar Mandal, University of Kalyani, India
Prof. Dr. Devadatta Sinha, Professor of CSE, University of Calcutta, India
Dr. Dharm Singh, Professor, Computer Science, Namibia University of Science and Technology, Namibia

Technical Programme Committee Co chair

Dr. Joyanta Kr. Roy, Technical Consultant (Automation), GKW Consult GmbH, India

Committee Members

Dr. S. K. Muttoio, University of Delhi, India
Mr. P. K. Hazra, University of Delhi, India
Prof. Vasudha Bhatnagar, University of Delhi, India
Prof. Sajal Saja, Kaziranga University, Jorhat, India
Dr. Partha Pratim Bhattacharya, Mody University, India
Dr. Sibaram Khara, Galgotias University, Noida, India
Dr. Santi Prasad Maity, IISET (BESU), Shibpur, India
Dr. P. K. Jana, IIT(ISM) Dhanbad, India
Dr. Debaprasad Mukherjee, Dr. B.C. Roy Engineering College, Durgapur, India
Dr. Saradindu Panda, Narula Institute of Technology, Kolkata, India
Dr. Raju Dutta, University of Calcutta, India
Dr. Bikramjit Sarkar, JIS College of Engineering, Kalyani, India
Dr. Arunava Dey, Professor, K.L. University, Guntur (AP), India
Dr Anirban Ghatak, Narula Institute of Technology, Kolkata, India
Dr. Punam Bedi, University of Delhi, India
Dr. Neelima Gupta, University of Delhi, India
Dr. Naveen Kumar, University of Delhi, India
S. Arun Kumar, IIT Delhi, India
Amitabha Bagchi, IIT Delhi, India
M. Balakrishnan, IIT Delhi, India
Subhashis Banerjee, IIT Delhi, India
Sorav Bansal, IIT Delhi, India
Naveen Garg, IIT Delhi, India
Rahul Garg, IIT Delhi, India
Ragesh Jaiswal, IIT Delhi, India
Prem Kalra, IIT Delhi, India
Saroj Kaushik, IIT Delhi, India
Amit Kumar, IIT Delhi, India
Anshul Kumar, IIT Delhi, India
Subodh Kumar, IIT Delhi, India
Mausam, IIT Delhi, India
Preeti R. Panda, IIT Delhi, India

Prof. (Dr.) M. Biswas, Jadavpur University, India
Prof. (Dr.) A. Choudhury, Jadavpur University, India
Prof. (Dr.) S. Mandal, NIT Durgapur, India
Prof. (Dr.) P. P. Sarkar, Kalyani University, India
Prof. Dr. Sourangshu Mukhopadhyay, University of Burdwan, India
Dr. Athitheyan, DG, MEDCOS DRDO, New Delhi, India
Dr. Arindam Dasgupta, DRDO, Hyderabad
Mr. Sandesh B. J., PESIT South Campus, India
Dr. Sudarshan T. S. B., PESIT South Campus, India
Dr. Annapurna, PESIT South Campus, India
Dr. Gowri Srinivasa, PESIT South Campus, India
Dr. Sarasvathi V., PESIT South Campus, India
Mrs. Pooja Agarwal, PESIT South Campus, India
Mr. Sajeevan K., PESIT South Campus, India
Kanthimathi S., PESIT South Campus, India
Mr. K. S. V. Krishna Srikanth, PESIT South Campus, India
Mrs. Keerti G. Torvi, PESIT South Campus, India
Ms. Bhuvaneswari K. J., PESIT South Campus, India
Ms. Sudeepa Roy Dey, PESIT South Campus, India
Ms. Sai Prasanna M. S., PESIT South Campus, India
Ms. Swati S. Gambhire, PESIT South Campus, India
Mrs. Neeta Ann Jacob, PESIT South Campus, India
Mrs. Shubha Raj K. B., PESIT South Campus, India
Mrs. Jermin Jeaunita T. C., PESIT South Campus, India
Mrs. Shanthala P.T., PESIT South Campus, India
Mr. Hanumanth Pujar, PESIT South Campus, India
Mrs. Sangeetha R., PESIT South Campus, India
Mrs. Surabhi Agrawal, PESIT South Campus, India
Mrs. Vandana M. Ladwani, PESIT South Campus, India
Mrs. Bidisha Goswami, PESIT South Campus, India
Mrs. Sudha.Y., PESIT South Campus, India
Mrs. Preethi Sangamesh, PESIT South Campus, India
Mrs. Rakhi Mittal Rathor, PESIT South Campus, India
Mrs. Ciji K. R., PESIT South Campus, India
Prof. Chittaranjan Hota, BITS Pilani, Hyderabad, India
Prof. R. Gururaj, BITS Pilani, Hyderabad, India
Prof. N. L. Bhanu Murthy, BITS Pilani, Hyderabad, India
Prof. Anand M. Narasimhamurthy, BITS Pilani, Hyderabad, India
Prof. Tathagata Ray, BITS Pilani, Hyderabad, India
Dr. G. Geethakumari, BITS Pilani, Hyderabad, India
Dr. Aruna Malapati, BITS Pilani, Hyderabad, India
Dr. Barsha Mitra, BITS Pilani, Hyderabad, India
Dr. Suvadip Batabyal, BITS Pilani, Hyderabad, India
Dr. Subhrakanta Panda, BITS Pilani, Hyderabad, India
Dr. Sudeepta Mishra, BITS Pilani, Hyderabad, India

Ms. Neha Bharill, BITS Pilani, Hyderabad, India
Mr. Gokul Kannan Sadasivam, BITS Pilani, Hyderabad, India
Shweta Agrawal, IIT Madras, India
Sutanu Chakraborti, IIT Madras, India
Timothy A. Gonsalves, IIT Madras, India
D. Janakiram, IIT Madras, India
Anurag Mittal, IIT Madras, India
Hema A. Murthy, IIT Madras, India
V. Krishna Nandivada, IIT Madras, India
Meghana Nasre, IIT Madras, India

Registration Chairs

Mr. Asit Mukherjee, Students' Welfare Officer, Kingston Polytechnic College,
  India
Mr. Arun Kr. Dasgupta, Principal in-charge, Kingston School of Management &
  Science, India
Dr. Kartik Kumar Kundu, Principal, Kingston College of Science
Smt. Sikha Mukhopadhyay, Director, Kingston Model School

Publicity Chairs

Dr. Suman Gupta Sharma, Principal, Kingston Law College, India
Mr. Tapan Roychowdhury, Treasurer, Kingston Educational Institute, India
Mr. Saugata Chakraborty, Training Placement Officer, Kingston Educational
  Institute, India

Organizing Committee Members

Dr. P. K. Chottopadhyay, Former Professor, Jadavpur University; Mentor, Kingston
  Polytechnic College, India
Prof. Dipak Kr. Bandyopadhyay, Former Professor, Jadavpur University; Mentor,
  Kingston Polytechnic College, India
Dr. B. P. Chottopadhyay, Former Professor, Jadavpur University; Mentor, Kingston
  Polytechnic College, India
Ms. Katia Routh, Kingston Educational Institute, India
Mr. Timir Chakraborty, Kingston Educational Institute, India
Mr. Dhiman Basu, Kingston Educational Institute, India
Mr. Utpal Dutta, Sasken Technology, India
Dr. Subhasis Sarkar, Kingston College of Science, India
Dr. Ipshita Haldar, Kingston College of Science, India
Mr. Deep Sadhu, Kalyani Government Engineering College, India
Mr. Arup Dutta, Kingston Educational Institute, India
Ms. Anwesha Nag, Kingston School of Management & Science, India
Ms. Shrabana Chottopadhyay, Kingston Law College, India
Mr. Sougata Sheet, University of Calcutta
Mr. Amit Saha, University of Calcutta
Mr. Mukul Banerjee, Kingston Educational Institute, India

Dr. Sumit Kumar Dey, Kingston College of Science, India
Mrs. Nibedita Kundu, Kingston Polytechnic College, India
Mr. Arvin Bera, Kingston Polytechnic College, India
Mr. Molay Kr. Giri, Kingston Polytechnic College, India
Mr. Sanjoy Kundu, Kingston Educational Institute, India
Ms. Jhilam Mukherjee, University of Calcutta
Mr. Abhik Ram Mondal, Kingston College of Science, India
Mrs. Ankita Marik, Kingston College of Science, India
Mr. Nilendu Chatterjee, Kingston College of Science, India
Mr. Partha Sarathi Kundu, Kingston College of Science, India
Krishnendu Guha, University of Calcutta
Mr. Prithwijit Chakroborty, Kingston Law College, India
Bulbul Roychowdhury, Kingston Law College, India
Ms. Riya Vijayan, Kingston Law College, India
Mr. Sandip Chanda, Kingston Law College, India
Mr. Tushar Kanti Dey, Kingston Law College, India
Ms. Mithu Mallick, Kingston Law College, India
Ms. Srabanti Kundu, Kingston Law College, India
Ms. Sumita Bhattacharjee, Kingston Law College, India
Mr. Nihar Ranjan Banarjee, Kingston School of Management & Science, India
Mr. Nirup Seal, Kingston School of Management & Science, India
Mr. Sumit Chakraborty, Kingston Educational Institute, India
Samadrita Chakraborty, Kingston Educational Institute, India
Ishani Dutta, Kingston Educational Institute, India
Babul Chakraborty, Kingston Educational Institute, India
Mr. Souvik Roy, Kingston Educational Institute, India
Syed Ariful Islam, Kingston School of Management & Science, India
Mr. Koushik Roy, Kingston Educational Institute, India
Mr. Raju Biswas, Kingston Educational Institute, India
Mr. Soumen Bera, Kingston Educational Institute, India
Mr. Subir Das, Kingston Educational Institute, India
Mrs. Barnali Bhattacharjee, Kingston Educational Institute, India
Ms. Chumki Biswas, Kingston Educational Institute, India
Mr. Surajit Paul, Kingston Educational Institute, India
Munmun Bhowmick, Kingston Educational Institute, India
Arup kr. Goswami, Kingston Model School, India
Ruma Bhattacharjee, Kingston Model School, India
Mahua Mitra, Kingston Model School, India
Sayonika Dhar, Kingston Model School, India
Abhijit Tripathy, Kingston Model School, India
Sumantra Dey, Kingston Model School, India

Hospitality Committee

Mr. Anirban Sarkar, Kingston Polytechnic College, India
Mrs. Sudipta Ghosh, Kingston Polytechnic College, India

Mr. Dipankar Chakraborty, Kingston Polytechnic College, India
Mr. Tapomoy Guha, Kingston Polytechnic College, India
Mr. Anower Hossain Gayen, Kingston Polytechnic College, India
Mr. Aurobinda Roy, Kingston Polytechnic College, India
Ms. Roneeta Purkayastha, Kingston Polytechnic College, India
Mr. Arunava Chakraborty, Kingston School of Management & Science, India
Mr. Arup Das, Kingston Polytechnic College, India
Mr. Maheswar Das, Kingston Polytechnic College, India
Mr. Mrinmoy Samanta, Kingston Educational Institute, India
Ms. Manashi De, Kingston Polytechnic College, India
Mr. Basudev Pal, Kingston Polytechnic College, India
Mr. Rabindranath Nath Jana, Kingston Polytechnic College, India
Ms. Soyeta Basak, Kingston Polytechnic College, India
Mr. Pritam Bhattacharyya, Kingston Polytechnic College, India
Mr. Tanmoy Biswas, Kingston Polytechnic College, India
Ms. Aratrika Maitra, Kingston Polytechnic College, India
Mr. Himadri Mitra, Kingston Polytechnic College, India
Mr. Nandadulal Chakraborty, Kingston Polytechnic College, India
Mr. Bikash Roy, Kingston Polytechnic College, India
Mr. Kartik Maity, Kingston Polytechnic College, India
Mr. Sauradeep Roy, Kingston Polytechnic College, India
Mr. Santu Maity, Kingston Polytechnic College, India
Mr. Debnarayan Chakraborty, Kingston Polytechnic College, India
Mr. Kiranmoy Samanta, Kingston Polytechnic College, India
Mr. Swapan Kr. Ghosh, Kingston Polytechnic College, India
Mr. Tanmoy Kundu, Kingston Polytechnic College, India
Mr. Liton Biswas, Kingston Polytechnic College, India
Mr. P. K. Basu, Kingston Polytechnic College, India
Mrs. Rupa Ganguly, Kingston Polytechnic College, India
Mr. Sourav Kundu, Kingston Polytechnic College, India
Mr. Souvik Mukherjee, Kingston Polytechnic College, India
Mr. Subha Mondal, Kingston Polytechnic College, India
Mrs. Debjani Bhattacharya, Kingston Polytechnic College, India
Mr. Papri Dhar Roy, Kingston Polytechnic College, India
Ms. Asmita Das, Kingston Polytechnic College, India
Ms. Priyanka Ghosh, Kingston Polytechnic College, India
Mr. Indodeep Chakraborty, Kingston Polytechnic College, India
Mr. Diptesh Ghosh, Kingston Polytechnic College, India
Mr. Harasit Basak, Kingston Polytechnic College, India
Mr. Sruti Sundar Bhowmick, Kingston Polytechnic College, India
Mr. Rominul Haque, Kingston Polytechnic College, India
Mr. Basudev Saha, Kingston Polytechnic College, India
Mr. Subhasis Maity, Kingston Polytechnic College, India
Mr. Titas Bhaumik, Kingston Polytechnic College, India
Mrs. Sutapa Biswas, Kingston Polytechnic College, India

Mr. Kaushik Nath, Kingston Polytechnic College, India
Ms. Jayeta Chakraborty, Kingston Polytechnic College, India
Mr. Saikat Chatterjee, Kingston Polytechnic College, India
Mr. Santu Sikdar, Kingston Polytechnic College, India
Ms. Pujayita Saha, Kingston Polytechnic College, India
Mr. Amit Kr. Das, Kingston Polytechnic College, India
Mr. Sourav Pal, Kingston Polytechnic College, India
Mr. Ramkamal Adhikary, Kingston Polytechnic College, India
Mrs. Debarati De, Kingston Polytechnic College, India
Mr. Subhasish Nath, Kingston Polytechnic College, India
Mr. Arunava Rano, Kingston Polytechnic College, India
Mr. Gopi Kanta Nath, Kingston Polytechnic College, India
Mr. Abhishek Roy, Kingston Polytechnic College, India
Ms. Ananya Das, Kingston Polytechnic College, India
Mr. Kalyan Mukherjee, Kingston Polytechnic College, India
Mr. Amar Kumar Samanta, Kingston Polytechnic College, India
Mr. Parimal Chandra Guha, Kingston Polytechnic College, India
Mr. Dibyendu Bairagi, Kingston Polytechnic College, India
Mr. Arnab Saha, Kingston Polytechnic College, India
Mr. Mainak Biswas, Kingston Polytechnic College, India
Mr. Sukriti Dan, Kingston Polytechnic College, India
Mr. Debkumar Das, Kingston Polytechnic College, India
Mr. Koushik Debnath, Kingston Polytechnic College, India
Ms. Prithagni Pal, Kingston Polytechnic College, India
Md. Tarik, Kingston Polytechnic College, India
Mr. Siddhartha Halder, Kingston Polytechnic College, India
Ms. Munmun Sarkar, Kingston Polytechnic College, India
Ms. Subhasree Poddar, Kingston Polytechnic College, India
Mr. Anup Sahoo, Kingston Polytechnic College, India
Mr. Subrata Roy, Kingston Polytechnic College, India
Mr. Ahasanul Haque Khan, Kingston Polytechnic College, India
Mr. Sreeparna Bhaumik, Kingston Polytechnic College, India
Mr. Sushmita Kanjilal, Kingston Polytechnic College, India
Mr. Pradip Kumar Pal, Kingston Polytechnic College, India
Mr. Debasis Bhattacharya, Kingston Polytechnic College, India
Ms. Tumpa Jana, Kingston Polytechnic College, India
Ms. Sudipa Hazari, Kingston Polytechnic College, India
Mr. Tanmoy Maji, Kingston Polytechnic College, India
Mr. Suvankar Dutta, Kingston Polytechnic College, India
Mr. Biswajit Das, Kingston Polytechnic College, India
Mr. Nirupom Som, Kingston Polytechnic College, India
Mrs. Mithu Saha, Kingston Polytechnic College, India
Mrs. Manika Bhattacharjee, Kingston Polytechnic College, India
Puja Sharma, Kingston Model School, India
Aparna Pal, Kingston Model School, India

Debi Biswas, Kingston Model School, India
Anindita Bagchi, Kingston Model School, India
Rupa Bachar, Kingston Model School, India
Soumendu Sahoo, Kingston Model School, India
Sanju Kumari Prajapati, Kingston Model School, India

Contents

Part III   Nature Inspired Computing

Part IV   Data Analytics

About the Editors

Prof. Jyotsna Kumar Mandal graduated from University of Burdwan with physics honours and did his post-graduation from Jadavpur University in physics and then M.Tech. in computer science and engineering from University of Calcutta in 1987. He did his Ph.D. in the field of computer science and engineering from Jadavpur University in 2000 and started his teaching career at North Bengal University. Currently, he is Professor at University of Kalyani. He was Visiting Professor at Vidyasagar University and Tripura University. He has also been a part of building academic programmes for different graduate and post-graduate levels at different universities. He has actively published in different journals and international conferences. He has also actively reviewed papers for many journals and conferences. He has guided research students for Ph.D. in computer science and engineering and M.Tech., B.Tech. and M.Sc. students for their dissertations. He was also in the editorial boards of different journals and conferences. He is an active member of different academic bodies in different institutions.

Prof. Dr. Devadatta Sinha graduated from Presidency College with mathematics honours and did his post-graduation in applied mathematics and then in computer science. He did his Ph.D. in the field of computer science from Jadavpur University in 1985. He started his teaching career in the Department of Computer Engineering at BIT Mesra, Ranchi, then at Jadavpur University and at Calcutta University from where he retired as Professor in the Department of Computer Science and Engineering. He also served as Head of the Department of Computer Science and Engineering. He also served as Convener, Ph.D. Committee in Computer Science and Engineering and in Information Technology, University of Calcutta. He also served as Vice-Chairman, Research Committee in Computer Science and Engineering, West Bengal University of Technology. During his career, he has written a number of research papers in national and international journals and conference proceedings. He has also written a number of expository articles in periodicals, books and monographs. His research interests include software engineering, parallel and distributed algorithms, bioinformatics, computational intelligence, computer education, mathematical ecology, networking. He has guided research students for their Ph.D. in computer science and engineering and M.Tech., B.Tech. and M.Sc. students for their dissertations. He has a total teaching/research experience of more than 38 years. He was also in the editorial boards of different journals and conference proceedings. He also served in different capacities in the programme committees and organizing committees of different national and international conferences. He was Sectional President, Section of Computer Science, Indian Science Congress Association, during the year 1993–1994. He is an active member of different academic bodies in different institutions. He is Fellow and Senior Life Member of CSI and has been involved in different activities including organization of different computer/IT courses since a long time. He is also adjudged Distinguished Speaker by Computer Society of India.

Prof. J. P. Bandyopadhyay is Former Professor and UGC Emeritus Fellow in the Institute of Radio Physics and Electronics, University of Calcutta. He is also Former Director of the Centre of Millimeter Wave Semiconductor Devices and Systems (CMSDS), University of Calcutta. He has a long and successful teaching and research career of about four decades in the Institute of Radio Physics and Electronics, University of Calcutta. His research interests cover semiconductor microwave and terahertz devices and photonic and quantum effect devices. He has published 184 research articles in peer-reviewed journals, several of which are highly cited. He has authored five textbooks in the field of electronics and telecommunication engineering for UG and PG students, some of which are recommended in the prescribed syllabus of various Indian universities as textbooks. He was the principal investigator of several R&D projects funded by DRDO, MIT, AICTE, CSIR, UGC, DST and DOE in the field of microwave and millimetre-wave semiconductor devices in the University of Calcutta. He has supervised the Ph.D. theses of a large number of students of Calcutta University. He was honoured several times to deliver invited talks in various research seminars, symposia and conferences held in India and abroad in the area of his teaching and research expertise. He has also chaired the technical sessions of various international conferences. He has been serving the cause of higher education, both teaching and research, by participating in Ph.D. committees and Board of higher studies as an external expert member of different premier institutes in India.

Part I
Computational Intelligence

A Hierarchical Image Cryptosystem Based on Visual Cryptography and Vector Quantization

Surya Sarathi Das, Kaushik Das Sharma,
Jayanta K. Chandra and J. N. Bera

1 Introduction

Today, the explosive growth in use of Internet is mainly involved in sharing of information which includes text data, audio, video, image, etc. Thus the confidentiality of information during transmission has turned into a major challenge nowadays. To avoid huge computation as in the standard encryption techniques like Advanced Encryption Standard (AES), Data Encryption Standard (DES), etc. Shamir and Naor [1] in 1994 proposed a simple image encryption technique for binary image called (k, n)-threshold visual cryptography scheme (VCS). As in [1], the basic (k, n)-threshold scheme visually encrypts a binary image into n number of images called shares. The security of the scheme is ensured by the condition that secret image is only recoverable if as a minimum k $(k \leq n)$ shares are available. When k numbers of shares or more are stacked, the visual system of a human helps to identify the secret image by looking at the stacked shares, without any mathematical computation. Later, VCS becomes popular in various application fields [2] and several researches can be found in the way to find VCS for grayscale images [3–5] and color images [6–8]. Another problem of pixel expansion in the basic formulation of VCS [1] instigated several researches [9–11] aiming to reduce pixel expansion. The shares generated in VCS are meaningless as it formed with some random pixel values. Thus it possesses a serious threat to be captured during

S. S. Das (✉)
Department of Computer Application, Kalyani Government Engineering College,
Kalyani 741235, India
e-mail: suryasarathi.das@gmail.com

K. Das Sharma · J. N. Bera
Department of Applied Physics, University of Calcutta, Kolkata 700009, India

J. K. Chandra
Department of Electrical Engineering, Ram Krishna Mahato Government Engineering
College, Purulia 723103, India

© Springer Nature Singapore Pte Ltd. 2019
J. K. Mandal et al. (eds.), *Contemporary Advances in Innovative and Applicable
Information Technology*, Advances in Intelligent Systems and Computing 812,
https://doi.org/10.1007/978-981-13-1540-4_1

transmission over any public network by curious hackers and secret image can be revealed by stacking them. Several researches [12, 13] may also be found to generate meaningful shares or innocent-looking shares to evade hackers. Thus, to devise a VCS that suits for (a) all kinds of images (b) with no pixel expansion and that (c) avoids the problem of meaningless share becomes a motivation of the research work presented in this paper. The authors of this present paper previously proposed a VCS employing a framework based on quantum signal processing [14]. The scheme in [14] works for the images of all kind viz. binary, grayscale and color images and does not involve any pixel expansion. The current paper is firstly focused to resolve the problem of meaningless shares. Hierarchically, in the first level, the QSP based VCS is used to generate shares. The meaningless shares are further encoded, in the next level, using a secret sharing technique derived from Shamir's scheme [15]. One of the pioneers of VCS, Adi Shamir [15], proposed in 1979, a (k, m) threshold scheme to share a secret based on a method of polynomial interpolation. The scheme in [15] divides any secret data into m parts, out of which minimum k parts are needed to rebuild the secret data. As the pixels of each share are encoded into m number of pieces, during transmission, even though hackers capture all the shares, it is not possible to reconstruct the secret image by stacking them as in VCS. But Shamir's scheme increases the data volume m-fold after encryption. This instigates the second objective of this present work to reduce the data volume overhead during transmission and thus to reduce the transmission delay. Vector quantization (VQ) [16] is a traditional quantization method mainly applied in signal processing, image processing and data compression. In the third level of the proposed scheme, VQ is used to lower the size of the encrypted data just before the transmission takes place. The complete block diagram of the proposed cryptosystem is shown in Fig. 1.

Thus the proposed scheme designs a image cryptosystems in its three-levels of hierarchy: (1) meaningless shares are generated using QSP based VCS, (2) to evade hackers, the meaningless shares are further encoded using Shamir's Scheme and finally (3) the encoded data volume is lowered using VQ to reduce transmission overhead.

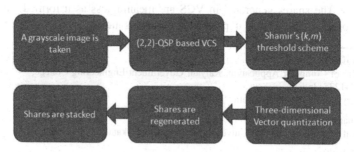

Fig. 1 The block diagram of the proposed cryptosystem

The paper is organized as: Sect. 2 describes proposed cryptosystem. The performance indices to evaluate the proposed scheme are described in Sect. 3. Results and performance analysis of the proposed cryptosystem are given in Sect. 4. Section 5 concludes the paper.

2 Proposed Cryptosystem

In this section, a detail step–by-step method is given to explain the proposed cryptosystem as follows:

Step 1: *(2, 2)-QSP based VCS of grayscale image*

The QSP based VCS [14] uses the notion of quantum mechanics to design a QSP framework. A QSP framework is designed with input mapping, then QSP measurement and finally output mapping. The framework processes an input image as a signal and applies the encryption algorithm to generate unexpanded shares. Here, in this paper, the simplest (2, 2)-QSP based VCS is used on a grayscale image I and two unexpanded shares S_1 and S_2 are generated. For the standard grayscale image "Lena", Fig. 2 shows both shares and the resultant image.

Step 2: *Employing Shamir's (k, m)-threshold scheme on each share*

In the first step of a (k, m)-threshold scheme, a polynomial function of degree $k - 1$ is chosen as follows:

$$f(x) = (a_0 + a_1 x + \cdots + a_{k-1} x^{k-1}) \bmod p, \tag{1}$$

where a_0 is the secret s to be secured and p is prime number large enough than s.

Then the value of the function f, say z, for m different values of x is calculated. These m numbers of z values correspond to the m different pieces in which the secret data s is encrypted. Out of these m pieces only k pieces suffice the reconstruction of the secret.

In the reconstruction phase, k pieces are randomly chosen. These k pieces are used to rebuild the coefficients of the polynomial function $f(x)$. Here, Lagrange

(a)   **(b)**   **(c)**   **(d)**

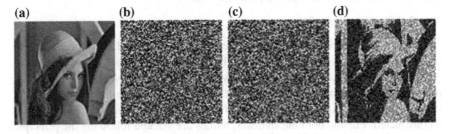

Fig. 2 QSP based (2, 2)-VCS: **a** Input image. **b** Share 1. **c** Share 2. **d** Resultant image

interpolation method is used find the Lagrange basis polynomial and the formula used is as follows:

$$l_i = \prod_{j=1, i \neq j}^{k} \frac{x - x_j}{x_i - x_j} \qquad (2)$$

Thus the polynomial function is reconstructed as:

$$f(x) = \sum_{i=1}^{k} (l_i \times z_i) \bmod p \qquad (3)$$

And finally, the secret s is retrieved by calculating the value of $f(0)$.

In this present paper, as an instance, the value of k and m are taken as 3 and 6 respectively to encrypt each of the pixel values for both the shares S_1 and S_2. It may be noted that, the size of each share S_1 (or S_2) generated after applying (2, 2)-QSP based VCS is same as the original image I, say $M \times N$. Thus after applying Shamir's scheme, two sets of data E_1 and E_2 of size $M \times N \times 6$ are generated corresponding to two shares. Obviously, this step increases the volume of encrypted data in a much greater scale.

Step 3: *Compression using Vector Quantization*

Vector quantization [16] is traditional quantization technique mainly used in signal processing and image processing. VQ takes a large set of input values and divides into groups. For an n-dimensional VQ, an n-dimensional grid pattern where the centroid of each cell is represented by n values is considered. Accordingly, n consecutive input values are considered at a time to map into index having nearest centroid value.

Here, three-dimensional VQ is used to encode E_1 and E_2 of size $M \times N \times 6$. Thus it reduces the volume of the encoded data by one-third. The codebooks C_1 and C_2, corresponding to two shares, generated after VQ, are transmitted to the receiver.

Step 4: *Decoding of codebook*

At the receiving end, the received codebooks C_1 and C_2 are decoded according to the vector de-quantization process. This decoding results in regeneration of E_1 and E_2 of size $M \times N \times 6$ at the receiving end.

Step 5: *Regeneration of shares*

In this phase, shares S_1 and S_2 are regenerated from E_1 and E_2 respectively. As because (3, 6)-threshold scheme is used here, only three values corresponding to each pixel from E_1 (and E_2) are randomly chosen. For each pixel, Lagrange basis polynomials are computed, polynomial function $f(x)$ is determined and pixel is regenerated. Then, after retrieving all $M \times N$ pixels, the share S_1 (and S_2) are reconstructed. Finally, both the shares S_1 and S_2 are stacked following the principle of VCS and secret image is regenerated.

3 Performance Indices

Following performance indices are used to evaluate the performance of the proposed cryptosystem:

Correlation Coefficient

The quality of resultant image is calculated here using correlation coefficient (*CC*) with respect to input image. The resultant image becomes qualitatively better when the value of the correlation coefficient is higher.

Information Entropy

How much the system is efficient to defy the security attacks can be measured by a parameter named Information entropy. Information entropy, introduced by Shannon in 1949, of an image represents the distribution of intensity levels. A truly random distribution of encrypted data is indicated by the entropy value near to eight.

Image encryption quality

It is a metric to determine the figure of merit of an image encryption technique. When an encryption technique is applied to an image, changes take place in pixels due to encryption process. The irregularity of such change is the measure of Encryption Quality. Higher value of Encryption Quality indicates greater quality of the image encryption technique.

Correlation between adjacent pixels

The correlation amongst the pixels in any image signifies the predictability of the image content. Low correlation value between adjacent pixels indicates random behavior of image and hence become unable to predict. The pair of adjacent pixels may be chosen from any orientation-horizontal, vertical or diagonal.

NPCR and UACI: parameters to evaluate the influence of differential attack

A good cryptosystem must be very sensitive to a small change made to the original image. An attacker may change plain image slightly, preferably one pixel change, to observe the changes in the encrypted image and thus to find a meaningful transition from original image to encrypted image. The influence of this differential attack can be quantified and parameters used for that are: Number of Pixel Change Rate (NPCR) and Unified Average Changing Intensity (UACI). NPCR measures the rate of change of number of pixels in encrypted image and UACI finds the average of intensity differences between two encrypted images. Higher value of NPCR indicates better randomization in the pixel distribution of the encrypted image.

4   Results and Performance Analysis

The cryptosystem proposed here is applied on three grayscale benchmark images-
"Lena", "Mandrill", and "Woman darkhair" [17], each of size 100×100 and
results are shown in Table 1. The comparative analysis of correlation coefficient
and values of Information Entropy, Encryption Quality, NPCR and UACI are listed
in Table 2. Table 3 lists the original image size, size of encoded data after Shamir's
scheme applied, size of the codebook. Table 3 also lists the percentage of change in
size amongst them. Table 4 lists the correlation coefficient between adjacent pixels.

Table 1 Results obtained

Original Image	Share # 1	Share # 2	Resultant Image

Table 2 Performance and security analysis

Image name	Correlation coefficient		Information entropy	Image encryption quality	NPCR	UACI
	Method as in [1]	Proposed scheme				
Lena	0.7158	0.7184	7.9522	73.6641	89.7400	33.5308
Mandrill	0.6890	0.6966	7.9807	73.4961	89.8700	33.0681
Woman darkhair [17]	0.7554	0.7622	7.9536	73.6909	89.0900	32.8129

Table 3 Changes in size

Image name	Size of each share generated after QSP based VCS applied (A) (KB)	Size of the encoded data for each share after Shamir's scheme applied (B) (KB)	Size of the codebook for each share (C) (KB)	Change between A and B	Change between B and C	Change between A and C
Lena	37	210	75	567.56% increase	35.71% decrease	202.70% increase
Mandrill	37	209	75	564.86% increase	35.88% decrease	202.70% increase
Woman darkhair [17]	34	209	75	617.70% increase	35.88% decrease	220.58% increase

Table 4 Correlation coefficient between adjacent pixels

Image name	Correlation between adjacent pixels					
	Horizontal		Vertical		Diagonal	
	Original	Encrypted	Original	Encrypted	Original	Encrypted
Lena	0.9036	0.1326	0.9036	0.1326	0.9036	0.1326
Mandrill	0.8164	0.0651	0.8164	0.0651	0.8164	0.0651
Woman darkhair [17]	0.9562	0.0118	0.9562	0.0118	0.9562	0.0118

As evident from the Table 2, the correlation value between the resultant image and original image are equally comparable in both cases, i.e., when only VCS in [1] is applied and when VCS followed by Shamir's scheme and VQ is applied. Thus, it can be asserted that the proposed cryptosystem does not have any negative effect when quality of the resultant image is concerned. Table 3 shows that when (3, 6)-Shamir's scheme is applied, size of the encoded data is increased by five to six times. When three-dimensional VQ is applied on the encoded data, the size increment is only limited to approximately two times of each share. The information entropy value, as shown in Table 2, which is almost 8 in all cases, affirms that the proposed cryptosystem maintains the desired level of randomness. The value of correlation coefficient between adjacent pixels in encrypted image, as shown in Table 4, is very low, in most cases it is near to zero. This result again strengthens the property of truly randomness of the encrypted share. Nonetheless, the Encryption quality of the proposed cryptosystem, as shown in Table 2, is found to be pretty high than AES where the Encryption quality value is 69.5250. All these findings assert that encrypted image does not bear any clue of the input secret image. Table 2 shows that the proposed system has value of NPCR of almost 90% and has value of UACI of almost 33%. Both these values ensure that the proposed cryptosystem is safe from differential attack.

5 Conclusion

The proposed cryptosystem successfully eliminates the problem of meaningless shares of VCS and removes the threat that if shares are captured by curious hackers and secret image is revealed by stacking the shares. Though the incorporation of Shamir's (k, m)-secret sharing scheme into QSP based VCS increases the size of the encoded data very high, the three-dimensional VQ successfully reduces the size of encoded in a manageable volume. The randomness of regenerated share images is still maintained in the desired level, as validated by the values of Information Entropy, Image Encryption Quality and Correlation between adjacent pixels. The chance of predictability of encrypted image is also safeguarded by the values of NPCR and UACI. Thus the proposed cryptosystem based on QSP based VCS, Shamir's secret sharing scheme and VQ strongly establishes a secure and simple image encryption technique.

References

1. Naor, M., Shamir, A.: Visual cryptography. In: Advances in Cryptography: Eurocrypt'94, pp. 1–12. Lecture Notes in Computer Science, Springer (1995)
2. Jana B., Samanta A., Giri D.: Hierarchical visual secret sharing scheme using steganography. In: Mohapatra R., Chowdhury D., Giri D. (eds) Mathematics and Computing. Springer Proceedings in Mathematics & Statistics, vol 139. Springer, New Delhi (2015)
3. Lin, C.C., Tsai, W.H.: Visual cryptography for gray-level images by dithering techniques. Pattern Recogn. Lett. **24**, 349–358 (2003)
4. Blundo, C., Santis, A.D., Naor, M.: Visual cryptography for grey level images. Inf. Process. Lett. **75**, 255–259 (2000)
5. Patel T., Srivastava R.: Hierarchical visual cryptography for grayscale image. In: Online International Conference on Green Engineering and Technologies (IC-GET), India, 2016
6. Hou, Y.C.: Visual cryptography for color images. Pattern Recogn. **36**, 1619–1629 (2003)
7. Liu, F., Wu, C.K., Lin, X.J.: Color visual cryptography schemes. IET Inf. Secur. **2**, 151–165 (2008)
8. Shankar, K., Eswaran, P.: RGB based multiple share creation in visual cryptography with aid of elliptic curve cryptography. China Commun. **14**(2), 118–130 (2017)
9. Hou, Y.C., Quan, Z.Y.: Progressive visual cryptography with unexpanded shares. IEEE Trans. Circuits Syst. Video Technol. **21**(11), 1760–1764 (2010)
10. Lin T.L., Horng S.J., Lee K.H., Chiu P.L., Kao T.W., Chen Y.H., Run R.S., Lai J.L., Chen R. J.: A novel visual secret sharing scheme for multiple secrets without pixel expansion. Expert Syst. Appl. **37**, 7858–7869 (2010); Elsevier
11. Askari, N., Heys, H.M., Moloney, C.R.: Novel visual cryptography schemes without pixel expansion for Halftone images. Can. J. Electr. Comput. Eng. **37**(3), 168–177 (2014)
12. Nakajima, M., Yamaguchi, Y.: Extended visual cryptography for natural images. J. WSCG2 **10**, 303–310 (2002)
13. Lee, K.H., Chiu, P.L.: Digital image sharing by diverse image media. IEEE Trans. Inf. Forensics Secur. **9**(1), 88–98 (2014)
14. Das S.S., Das Sharma K., Chandra J.K., Bera J.N.: Quantum signal processing-based visual cryptography with unexpanded shares. J. Electr. Imaging, SPIE **24**(5), 053026-1–053026-18 (2015)

15. Shamir, A.: How to share a secret. Commun. ACM **22**(11), 612–613 (1979)
16. Orest V.O., Mircea W.: Improving vector quantization in image compression with Hilbert scan. In: 20th International Conference on Systems, Signals and Image Processing (IWSSIP), Buchtarest, Romania (2013)
17. http://www.imageprocessingplace.com/downloads_V3/root_downloads/image_databases/standard_test_images.zip. Accessed 7 June 2012

Intelligent Web Service Searching Using Inverted Index

Sinthia Roy, Arijit Banerjee, Partha Ghosh, Amlan Chatterjee
and Soumya Sen

1 Introduction

The OASIS standard protocol Universal Description, Discovery and Integration (UDDI) [1] is used to define a protocol for the underlying networks in a Service-Oriented Architecture (SOA) [2] platform to publish and discover software components. It is defined in UDDI registry where specific Web Services [3] are offered and who provide these by defining associated APIs and data structures to publish service descriptions in the registry.

As the new web services are being deployed the magnitude and volume of UDDI Business Registries (UBRs) are growing rapidly across the internet. Obviously managing this high usage and size of the UDDI Business Registries is a challenge and therefore suitable search mechanism is required. This could be considered both from technical and business logic perspective. The basic search methods are offered

S. Roy
Guru Nanak Institute of Technology, Kolkata, India
e-mail: roy.sinthia@gmail.com

A. Banerjee
Cognizant Technology Solutions, Kolkata, India
e-mail: info.arijit@gmail.com

P. Ghosh
Department of Computer Application, Kingston School
of Management & Science, KEI, Kolkata, India
e-mail: pghosh44@gmail.com

A. Chatterjee
California State University, Dominguez Hills, Carson, USA
e-mail: achatterjee@csudh.edu

S. Sen (✉)
A.K. Choudhury School of IT, University of Calcutta, Kolkata, India
e-mail: iamsoumyasen@gmail.com

© Springer Nature Singapore Pte Ltd. 2019
J. K. Mandal et al. (eds.), *Contemporary Advances in Innovative and Applicable Information Technology*, Advances in Intelligent Systems and Computing 812,
https://doi.org/10.1007/978-981-13-1540-4_2

by present UDDI to find out the appropriate web services from the available multiple UDDI Business Registries. These searching times are to be reduced in real-time systems. It is even more challenging when the services are accessed through devices having lower computing capabilities such as mobile devices and others. This basic but crucial requirement motivates the researchers to look for new techniques that allow faster retrieval of business logic and corresponding web services. Search engines [4] are helpful in this context to identify locate the appropriate information from the registry.

In this research work the searching facility is enhanced by incorporating search engine in UDDI registry. Initially service providers publish their one or more web services in the UDDI registry. The updated registry information is stored as WSDL document [5], which contains the following entities: *businessEntity*, *businessService, binding template* and *tModel*. API calls and SOAP messages [6] are used to transfer information between the service provider and the registry. It is followed by the formation of index database using the <description> label of the *businessService* and *businessEntity*. These two separate index databases are generated using the concept of inverted index. Next during the discovery phase, the matching of search query is done with the keywords defined in the index database. Now based on the corresponding key matching the keywords are returned to the users. The accessing behavior of the current users is stored in log files, which contain two separate files for *businessService* and *businessEntity*. Once, more and more users start to use this search engine facility, the amount of records in the log file obviously increase and this allow to offers the customized business services to the users by analyzing log files. New web services are also being added by service composition [7] on the existing web services.

In Sect. 2 proposed system is described along with its architecture, system functionality and in Sect. 3 conclusion and future work is discussed.

2 Proposed System

The discovery mechanism of the existing system primarily concentrates on comparing keywords in service customer's query. Therefore these techniques may result in irrelevant search, as new web services are published in the registry with same keywords. In order to address this problem we propose a two phase search engine that reduces the time requirement to discover web services and also offers the users relevant web services.

In this research work existing UDDI is extended based on search engine. In order to create the intelligent search engine, we have conceptualized two terminologies.

1. Registry Index Database (RID)
2. User Orientation Database (UOD).

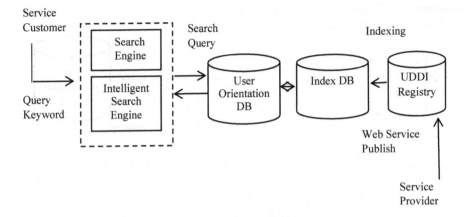

Fig. 1 Proposed system architecture

Initially, the system will use the index database to return web service to service customers. Over a period of time as more and more queries hit the system, user orientation database will play a significant role. We need to choose the key word based on the existing business knowledge, as we have generic idea about the user requirements. Across a span of time as more queries hit the system we can apply our own searching mechanism to identify the keys (Fig. 1).

In the proposed system search keywords are received from the service user through the search engine user interface. The search query is then sent to the user preference database. User orientation database contains log files, which in turn has the web services grouped under several service group identifiers. Service group identifiers club the different web services based on the business they implement. The search keyword is then compared to service group identifier in the log file. The web services in the respective group are then returned to the entity Service Customer. In case the required Service Group Identifier is not found in the user preference log file, the search keyword is sent to the Index Database. At the index database, the web services which are published at the UDDI are indexed. Indexing will help retrieval of suitable web services faster and also improve the performance web searching. The index database is queried with the search keyword. The service key(s) are stored against keywords in index database. These keywords are actually the service group identifiers. The user query keyword is forwarded to the index database by the user preference database. Now user query keyword is compared with the keyword at the index database. The row which matches the search, corresponding *businessKeys* or *serviceKeys* is returned to the user orientation database. At the user orientation database, the user query keyword is entitled to be the service group identifier. Under this service group identifier, those *businessKeys* or *serviceKeys* are clubbed and saved. Next time, if any user comes in searching for this kind of web service, the user preference database would be well equipped to return the data without the help of the index database. This would save the time of

```
select * from dbo.tbl_ServiceEntries
where keywords like 'Restaurant'
```

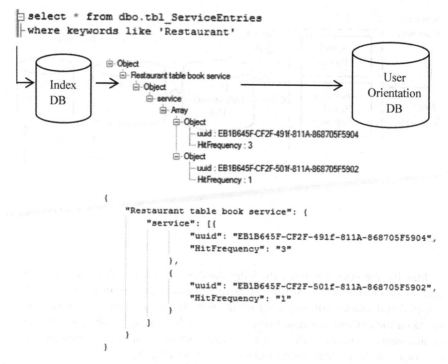

```
{
    "Restaurant table book service": {
        "service": [{
            "uuid": "EB1B645F-CF2F-491f-811A-868705F5904",
            "HitFrequency": "3"
        },
        {
            "uuid": "EB1B645F-CF2F-501f-811A-868705F5902",
            "HitFrequency": "1"
        }
        ]
    }
}
```

Fig. 2 Transformation of relational data to non-relational data

doing SQL transaction at the index database and doing away with fetching data from the normalized tables by several join operations. The idea is depicted in Fig. 2.

At first place, the web services which are published at the UDDI registry would be indexed at the Index database.

2.1 Preparing Indexed Registry Database—Web Service Indexing

After the service providers register their services at UDDI, the web services needs to be indexed in order to retrieve the web services in an optimized approach. Indexing will help in comfortably fetching the right web service(s) under the right business group (service group identifier). Once the web services are registered at the UDDI, the index database takes shape instinctively.

There are many indexing methods like, Suffice tree index, Citation Index [8], Inverted index [9], Ngram index [10] and Document-term index [11].

In context to the scenario of the proposed system we reckon that inverted index fits appropriately. The reason would be, assuming the web services are registered at the UDDI, the description element from all the *businessEntity* documents for the registered web services are taken into account. From these descriptions, unique description would be identified as keyword, and all the web services that have the same keyword, in their respective description element, would be clubbed under the unique keyword. In this way the index will be formed. Since this kind of index is generated from the existing documents, this is termed as inverted index. As this is generated dynamically based on the actual content it is used over other indexing technologies in this context.

Therefore for published web services inverted index is preferred here. Inverted index is primarily used to collect a mapping from the content, for example words to its position in a database or in a set of documents.

Here we propose to have two tables for index database making use of inverted index, IB for *businessEntity* and IS for *businessService*. IB is built from the explanation given by the service provider in the *businessEntity* while registering the business in UDDI. IB schema primarily contains two columns namely, Keyword and *BusinessKeys*. Keywords column would have the unique keyword from the description element from the *businessEntity* documents of the registered web services. *BusinessKey* would have the unique *businessKey* from the *businessEntity* document. Likewise, It is built from the information given by the service provider in *businessService* while registering the service in UDDI.

IS schema primarily contains two columns namely; Keyword and *ServiceKeys* (refer to Fig. 3). Keywords column would have the unique keyword from the description element from the *businessService* documents of the registered web services. *ServiceKey* would have the unique *serviceKey* from the *businessService* document (refer to Fig. 4).

Once a new service provider arrives, the probable keywords are identified. Now these business keys associated with these keywords are included in the index database along with these keywords. If the keyword is already in the index

Fig. 3 Inverted index for *businessEntity* (IB table)

Keywords	Business Keys
Hotel	Key1,key2,key3,key4
Air travels	key1, key2, key3
Hospital	key1, key2
Reservation	key1, key2, key3, key4

Fig. 4 Inverted index for *businessService* (IS table)

Keywords	Service Keys
Food delivery	Key1,key2,key3,key4
Hospitality	key1, key2, key3
Lodging	key1, key2
Internet	key1, key2, key3, key4

database, the *businesskey* is appended to the Business key field that matches to the keyword.

In this way the index database is created. We would now go ahead and search a web service using the Index database.

2.2 Searching Web Service Using Index DB

This is the phase when the system is newly introduced and it has been hardly used by any user. In this scenario, when a user searches for a web service and gives a keyword, it hits the index database. The keyword given by the user is then taken up and compared with the "Keyword" in the index database. The row which meets the condition is returned which contains the corresponding unique keys.

If the user searches a service related to a kind of business, then the search process will hit the IB table. The row which has the same keyword like the search query is returned, which has the corresponding unique business keys. Now, this business key is used to search the relevant web services from the IS table. Eventually, list of these web services' unique identifiers are returned to the end user.

Index database is now active and is capable of returning relevant results against all the queries that would hit the index database. But yet there is some work to be done in context to analyze users' behavior when it comes to searching for a web service. If one's behavior can be categorized then it would yield better results in terms of returning relevant web services. This is the reason we move on to user preference database, as the name says, which works only on user preferences.

2.3 Inception of User Orientation Database in Intelligent Search Engine

The major motivation of an intelligent web search engine is to identify customized or personalized web services by analyzing the users' searching behavior in order to retrieve the web services which might be aligned to a specific kind of business or a certain kind of service.

In order to achieve this, we would like to bring in a User Orientation database which will consist of a log file. The log file is used to store to behavior of users all along the life of the system. By analyzing this log file we plan to render personalized, more relevant and frequently used web services to the end users.

2.4   User Orientation Database

User orientation database is entitled to get the hit at first. Initially when the log will be empty, flow will be redirected to the index database. When a search keyword arrives to the user orientation DB, the log files will be searched at first.

We propose these log files to be residing in a NoSQL DB. So, that will facilitate us to add or remove a web service entry against a keyword if it is most frequently used or less frequently used respectively. Moreover the advantage of NoSQL DB being used as a cache, we need not maintain any relational schema. With relational database as a cache, there could have been multiple join operations for listing a single web service. But with NoSQL database there would be no such scenario. NoSQL database also returns data in a tree format so we need not parse the data every time unlike the data from a relational database. We would not like to parse a data which is being frequently used by the service customers.

Most frequently searched data will be fetched from index DB and stored in user orientation DB. These data will be stored in user preference DB as long as those are relevant and returned to the users. When they would become irrelevant, they would be moved out and instead the services which are more relevant will take their place in user orientation DB.

User orientation DB will contain two log files, a business log file and other kind will be the service log file.

Referring to the business log file for an instance when a business will be searched, "hotel reservation service" would be the search keyword in this case. As per the proposed system it will be first searched in user orientation DB. If the keyword exists in user orientation DB then corresponding UUID/business key of the business service will be returned based on its frequency of hit. Higher the frequency more is the relevance.

Service log file will have the schema something like the following (Fig. 5).

```
{
    "food delivery service": {
        "service": {
            "BusinessUuid": "EB1B645F-CF2F-491f-811A-868705F5904",
            "ServiceUuid": "ESDG645F-CF2F-491f-911A-88905F5904",
            "HitFrequency": "3"
        },
        {
            "BusinessUuid": "EB1B645F-CF2F-491f-811A-868705F5904",
            "ServiceUuid": "EB1B645F-CF2F-491f-811A-868705F5902",
            "HitFrequency": "1"
        }
    }
}
```

Fig. 5 Service log file schema

At any instance when the user is looking for a specific web service rather than an entire business implementation, which obviously would belong to a certain business, flow will first come to user preference DB. User orientation DB will be queried with the search keyword. If any record matching the keyword happens to be present then the web service having the maximum hit frequency will be returned.

2.5 Grouping of Web Services Based on Business

In service log file of a travel system user query contains the categories like hotel booking service, food options service, internet facility service etc. With the motivation to find the most relevant service, we propose to group the set of records (shown in the above figure against "hotel reservation service" in this case) based on the same search query ("hotel reservation service" in this occasion). Using the user query field array of records having the same user query is kept as an individual group. This helps us to trace the service easily. For example, let us consider an occasion where the user query is "hotel reservation service". The records that are entered against the user query "hotel reservation service" in the index database are kept under one group in user preference database keeping a count of the number of times they are returned to user.

2.6 Service Discovery Using User Orientation DB

When the system will be live for a while, the user orientation DB will be of more use than it used to be earlier. If a new user looking for a web service, user is required to key in the search keyword (i.e., the user query). The user query is then sent to the user preference DB layer. The user orientation DB is searched tallying the user query. The group which matches the user query is identified. Based on the frequency of each of the records in each group, the one having the highest among all is returned to user.

3 Conclusion and Future Work

We have proposed UDDI based intelligent search engine extending the standard UDDI primarily to discover a more relevant web service. The proposed work contains the conventional search engine and intelligent search engine on UDDI. Initially, web services are published in the registry. The published web services are indexed and maintained in the index database. The most relevant web service is offered to the user by making use of the search engine concept, besides accessing the records from the log files. The proposed system implements registry of web services simultaneously which is capable to offering effective and faster searching facility using an inverted index database along with a NoSQL database.

This research work could be extended further for different NoSQL databases. Specific properties of these databases could be exploited for better performance of the corresponding system. For example in MongoDB, shredding property may be integrated over the proposed methodology for effective implementation of web service indexing.

References

1. Fang, W., Moreau, L., Ananthakrishnan, R., Wilde, M., Foster, I.: Exposing UDDI service descriptions and their metadata annotations as WS-resources. In: Proceedings of the 7th IEEE/ACM International Conference on Grid Computing, pp. 128–135 (2006)
2. Sneed, H.M.: Migrating from legacy to SoA. In: IEEE 9th International Symposium on the Maintenance and Evolution of Service-Oriented and Cloud-Based Environments (2015)
3. Li, S., Wen, J., Luo, F., Gao, M., Zeng, J., Dong, Z.Y.: A new QoS-aware web service recommendation system based on contextual feature recognition at server-side. IEEE Trans. Netw. Serv. Manage. 14(2), 332–342 (2017)
4. Tewari, V., Dagdee, N., Singh, I., Garg, N., Soni, P.: An improved discovery engine for efficient and intelligent discovery of web service with publication facility. In: Proceedings of World Conference on Services—II, Bangalore, pp. 63–70, Sept 2009
5. Jiang, L., Liu, T., Liu, D.: Objective and SUbjective QoS factors supported web service search method based on extended WSDL. In: 23rd International Conference on Geoinformatics (2015)
6. Issac, S., Uma Devi, V.: Efficient querying and SOAP based streaming of multimedia content using web services. In: International Conference on Intelligent Computing Applications (2014)
7. Bhattacharya, A., Sen, S., Sarkar, A., Debnath, N.C.: Hierarchical graph based approach for service composition. In: Proceedings of the 17th IEEE International Conferences on Industrial Technology (ICIT 2016)
8. Day, R.E.: Representing documents and persons in information systems: library and information science and citation indexing and analysis. In: Indexing It All: The Subject in the Age of Documentation, Information, and Data. MIT Press eBook Chapters (2014)
9. Arab, A., Abrishami, S.: MDMP: a new algorithm to create inverted index files in BigData, using MapReduce. In: 7th International Conference on Computer and Knowledge Engineering (ICCKE) (2017)
10. Thacker, U., Pandey, M., Rautaray, S.S.: Performance of elastic search in cloud environment with nGram and non-nGram indexing. In: International Conference on Electrical, Electronics, and Optimization Techniques (ICEEOT) (2016)
11. Sorkun, M.C., Özbey, C.: Compression experiments on term-document index. In: International Conference on Computer Science and Engineering (UBMK) (2017)

Computational Intelligence Based Neural Session Key Generation on E-Health System for Ischemic Heart Disease Information Sharing

Arindam Sarkar, Joydeep Dey, Anirban Bhowmik,
Jyotsna Kumar Mandal and Sunil Karforma

1 Introduction

Cryptographic approach is best suited for transmitting any vital information in real world [1, 2]. In this advanced technological era, E-Health System is an emerging area which serves the public health care services. Community gets quick and secured treatment by utilizing the services provided by this medical domain. Data fabrication based on existing encryption techniques is being applied here [3–5]. In any medical system, there needs to communicate different types of patients' medical information. In order to achieve secured online transmission between different parties, ciphering is being done on the medical data. In case of Digital–Cardiological Systems, different cryptographic algorithms are being applied on the

A. Sarkar (✉)
Ramakrishna Mission Vidyamandira, Belur Math 711202, India
e-mail: arindam.vb@gmail.com

J. Dey
M.U.C. Women's College, B.C. Road, Bardhaman 713104, India
e-mail: joydeepmcabu@gmail.com

A. Bhowmik
Cyber Research and Training Institute, The University of Burdwan,
Bardhaman 713101, India
e-mail: animca2008@gmail.com

J. K. Mandal
Department of Computer Science and Engineering, University of Kalyani,
Kalyani 741235, India
e-mail: jkm.cse@gmail.com

S. Karforma
Department of Computer Science, The University of Burdwan,
Bardhaman 713104, India
e-mail: dr.sunilkarforma@gmail.com

© Springer Nature Singapore Pte Ltd. 2019
J. K. Mandal et al. (eds.), *Contemporary Advances in Innovative and Applicable Information Technology*, Advances in Intelligent Systems and Computing 812,
https://doi.org/10.1007/978-981-13-1540-4_3

cardiological data like ECG, Echocardiograph, Tread Mill Test, Holter Monitoring, etc. Patients, Cardiologists, Physicians, Pathologists, Hospital Administrators, etc., are the integral users of such digital systems. This paper presents a cryptographic approach towards transmitting an ECG signal revealing Ischemic Heart Disease from patients to Cardiologists. The most vulnerable issue while online transmitting such heart related information is the security implementations over the Internet. Heart is the foremost integral component of human body to remain alive. Lack of supply of blood to the heart is a diseased condition termed as Ischemic Heart Disease (IHD). Deposition of cholesterol on the walls of blood vessels catalyzes to be narrower or blocked by minimizing flow of supply of oxygen to the heart muscles [6]. This blockage mechanism affects the functioning of the heart. IHD is a set of cardiac-related diseases [7] like myocardial infarction, silent ischemia, stable or unstable angina, etc. Electro Cardio Graph [8] based on 12 lead is a preliminary diagnosis technique to detect IHD. ECG technique which contains P, Q, R, S, and T standard waves signals for depolarization and repolarization of the heart muscle. Analog ECG signal of an IHD patient is being converted into binary bits using Analog-to-Digital Converter device. A wide set of variations for secured transmission techniques are often used to abstract data from eavesdroppers [9–12]. Cryptographic algorithms have some limitations too. The key used for encryption, guides the level of security on the entire cardiological information. In such E-Health domain, intruders lying in the middle of the channel can easily steal the key value during the exchange of secret key via insecure public transmission channel [13, 14].

This paper presents a protected session key construction mechanism based on Two Tier Neural Network. A novel technique embedded with simultaneous key exchange and authentication approach has been proposed. Patient and cardiologist use the same secret common weight vector.

The organization of this paper is as follows. Section 2 deals with problem specification. Proposed technique has been explained in Sect. 3. Section 4 contains the experiments results. Analysis in the context of different aspects of the technique and results combined with transmission risks and security ensure policy have been illustrated in Sect. 5. Conclusions and further scopes of developments are discussed in Sect. 6 and references are at the end.

2 Problem Specification

2.1 Problem of Fake Cardiac Insurance Claim

By stealing the confidential information in the mid-way, intruders may adopt for unfair and fake Cardiac Insurance claim. Re-imbursement of treatment cost may be generated without consent of the patient, which is not ethical.

3 Proposed Technique

The above stated problem may be addressed in the proposed technique with no need to exchange a secret session key. Here, a symmetrical Neural Network model has been implemented by both the patient and cardiologist to generate the session key. The concept of identical Two-Tier Neural Network machine is being adopted. Both end parties use same input vector generated from Pseudorandom Number Generation to input both machines. Corresponding hidden layer nodes are being assigned with a random weight values. Such weights are generated by the predefined synaptic links of TTNN. Attackers have no idea about the internal architecture of both the machines and corresponding symmetrical input vector. Based on input vectors and randomly assigned weights value, both the machine produces some output values. Final outputs are being transmitted over public channel. In case of same output values generated by the both machines, then learning step is to be applied for synchronization. The synchronized patient's machine cardiologist's machine generates identical weight vectors termed as session key. Session key exchange procedure will be followed by key authentication technique in parallel fashion. Authentication is done by transmitting last m bits of the identical weight between patient and cardiologists. If both the sequences are identical, then only authentication succeeds, otherwise not. Attacker does not have identical weight vector. By sniffing the public channel, attacker can steal some bits without prior knowledge. Even if the attacker hacks the m bits, then to predict the remaining $(p - m)$ bits of the identical weight vector, he/she has to perform $(p - m)!$ combinations, that is computationally infeasible. Total no. of bits present in the identical weight vector is p. A neural network usually contains R no. of hidden neuron nodes, N number of input neuron nodes with binary input vector as $x_{ij} \in \{-1, +1\}$, and the discrete neuron weights as $W_{ij} \in \{-L, -L+1, -L+2, \ldots, +L\}$ where, $i = 1, 2, \ldots, R$ represents the ith hidden neuron, and $j = 1, 2, \ldots, N$ represents the elements of vector and an output neuron. Specific output of hidden layer neuron is computed using weighted sum value depending on present input. Thus, the state of any hidden neuron can be expressed by the following equation number 1.

$$h_i = \frac{1}{\sqrt{N}} \sum_{j=1}^{N} W_{i,j} x_{i,j} \tag{1}$$

The output of ith hidden neuron may be defined as $\partial_i = \text{SGN}(h_i)$.

But in case of $h_i = 0$ then $\partial_i = -1$, to produce a binary output. Hence, $\partial_i = +1$, if the weighted sum value is positive, otherwise, this is $\partial_i = -1$. Total output value of a network is the product of the hidden sub units computed by the following equation number 2.

$$\tau = \prod_{i=1}^{R} \partial_i \tag{2}$$

In this case, the value of R will be splitted into $R1$ and $R2$, where $R2$ denotes the number on hidden nodes just adjacent to the corresponding output layer. For each $R2$ neurons, there exists $R1$ number hidden neurons. Now for every $(R1 \times R2)$ neurons, there would be N possible inputs. Thus, the input internal layer has $(R1 \times R2 \times N)$ input neurons to denote the size of the Two Tier Neural Network. First hidden layer numbered as 1, i.e., with $(R1 \times R2)$ neurons, neuron generates ∂_i^1 values and second hidden layer numbered as 2, i.e., $R2$ neurons generates ∂_i^2 values, as presented below.

$$\partial_i^1 = \text{SGN} \left(\sum_{j=1}^{N} W_{i,j} x_{i,j} \right) \tag{3}$$

$$\partial_i^2 = \text{SGN} \left(\sum_{j=1}^{N} \partial_i^1 \right) \tag{4}$$

The function SGN returns -1, 0 or 1, defined in Eq. 5.

$$\text{SGN} = \begin{cases} -1 & \text{if } x \text{ less than } 0; \\ 0 & \text{if } x \text{ equals to } 0; \\ 1 & \text{if } x \text{ greater than } 0; \end{cases} \tag{5}$$

When the value of the scalar product is equivalent to zero, then output of the hidden unit is replaced by -1. The output of this double tier neural network is calculated as the product of all the values those are generated by hidden nodes as in Eq. 6.

$$\tau = \prod_{i=1}^{R2} \partial_i^2 \tag{6}$$

The learning procedure between the patient and cardiologist is explained as follows.

1. If the outputs, $\tau^P \neq \tau^C$, then learning not needed.
2. If $\tau^P = \tau^C = \tau$, only the weights of the hidden nodes having $\partial_R^{P/C} = \tau$ would be changed.
3. The corresponding weight vector of such hidden node is fine tuned by the following learning rule [2–4].
   Anti-Hebbian: $W_R^{P/C} = W_k^{P/C} - \tau^{P/C} x_R \, \theta(\partial_R \tau^{P/C})(\tau^P \tau^C)$.

In such Two-Tier Neural Network based key generation mechanism, if the patient and cardiologist do not get the identical/same input vectors, i.e.,

$\forall (tm): x^P(tm) \neq x^C(tm)$, then the synchronization is not gained. If the inputs are symmetrical for both parties, then only both machines are being trained using each other outputs. With diverse range of inputs, the two end systems try to learn totally unknown relations between inputs $x^{P/C}(tm)$ and $\tau^{P/C}(tm)$. It prevents the generation of such time dependent equal weight vectors. The sustainable developments of normalized sum value of the absolute differences DIFF $\{w^P(tm), w^C(tm)\} \in [0, 1]$ plotted over time for different offsets $\forall (tm): x^P(tm) = x^C(tm + \varphi)$, $\varphi \in N$ in the input vector with completely different unknown input vector. Now $W_{Rj}^P(tm)$ and $W_{Rj}^C(tm)$ get a unequal random element $x_{Rj}(tm)$ out of their input vectors. Therefore, the distance in between the any two elements will not to be condensed to zero value. After every bounding action, patient and cardiologist deviate. Hence, common input vector does not ensure synchronization of machines. Due to this cause, the common inputs, i.e., $x^{P/C}(tm)$ and $x^{C/P}(tm)$ are kept secret between the patient and cardiologist. Neural machines generating their individual assigned secret initial weights $W^{P/C}(tm)$ are made unknown to each other. Brute Force attacks become practically non viable and highly expensive due to $(2RN - 1)$ calculations are required to explore all possible valid common input combinations. Using such authentication procedure, attacks that occur in the format of Man-In-The-Middle attack and other fake mediclaim attacks may be restricted.

At the time of synchronization procedure, no information on the common secret key is seep out at all. The only information transmitted is the unknown bit-strings. In the case of the authentication, the inputs are randomly chosen only for certification rule. In real scenario, an attacker also cannot differentiate an authentication step from a synchronization step by observing the exchanged outputs. As the attackers do not know the common input vector, so they are unable to predict whether the currently observed output bit is used for authentication or synchronization purpose.

4 Experimental Results

It deals with the outcomes on different types of ECG information with different sizes with extensive performance analysis. Comparative study between the proposed technique and RSA algorithm, Triple-DES (168 bits), AES (128 bits) has been carried out on 20 files of ECG revealing Ischemic Heart Disease by doing different types of experiments. The ECG signals are taken from MIT-BIH ST Change Database [15] for experimental purpose. Frequency test is shown in Table 1. This is being done to find proportion of ones and zeros in the binary ECG sequence. It predicts whether the number of zeros and ones in such sequence are approximately the same as it is expected for random sequence. Table 2 shows distribution of frequency within a block of the ECG sequence of bits. It has been carried out to find the proportion of ones and zeros within $M1$ bit blocks. It determines whether the frequency of ones within a $M1$ bit block is approximately

Table 1 Proportion of successful uniformity of frequency distribution

Technique	Proportion of expectation	Proportion of observation	Status for proportion of success	P-value of P-values
Proposed	0.972766	0.985437	Success	4.102711e−10
TDES		0.983333	Success	3.571386e−01
AES		0.984871	Success	3.915294e−07
RSA		0.986667	Success	4.122711e−10

Table 2 Proportion of passing and uniformity of distribution for frequency within a block

Technique	Proportion of expectation	Proportion of observation	Status for proportion of success	P-value of P-values
Proposed	0.972766	0.986942	Success	3.929802e−01
TDES		0.980000	Success	3.639271e−06
AES		0.984792	Success	3.903719e−03
RSA		0.990000	Success	3.949802e−01

Table 3 Proportion of successful uniformity of distribution within a run

Technique	Proportion of expectation	Proportion of observation	Status for proportion of success	P-value of P-values
Proposed	0.972766	0.992163	Success	1.181790e−01
TDES		0.986997	Success	1.160128e−01
AES		0.990000	Success	1.174101e−01
RSA		0.993333	Success	1.191964e−01

$(M1 * 0.5)$. Table 3 shows the uniformity of distribution for runs. The aim of this test is to identify an uninterrupted sequence of ones and zeros runs in the entire ECG binary information. A k length run denotes that a run consists of exactly identical bits sequence in the entire string.

An inference may be drawn by observing Tables 1, 2 and 3 that the proposed technique along with existing classical techniques, have successfully passed the frequency test, frequency within a block test and runs test. The observed proportion values of the proposed technique are larger than expected proportion value. In short, it is also noted that proposed technique outperform than existing TDES and AES technique.

5 Analysis of Results

The proposed technique yields Chi-Square value which is greater than the TDES and good enough to compare with RSA algorithm. Thus, it may be ensured the degree of randomness and wide spread of the characters between the range of 0 and 255 parameter value. The experimental results define the optimal performance of the proposed technique. In this technique, the patient and cardiologist need not to transmit an identical secret key; instead they utilize their non-distinguishable weights as a common secret key for the encryption purpose. In Brute Force attack having R number of hidden neural neurons, $(R * N)$ inputs and boundary of weights L, produces $[(2L + 1) * (R * N)]$ possible combinations. For example, if the TTNN contains $R = 3$, $L = 3$ and $N = 100$, then it gives $(3 * 10{,}253)$ chances of key possible combinations, which guides the attack towards unfeasible even with high computational skills.

6 Conclusion and Future Scope

A secured novel cryptographic approach towards generation of secret key has been proposed using TTNN synchronized mechanism. It raises the security concerns of the secret key exchange algorithm by increase in the synaptic link depth of L inside the TTNN architecture. In this case, patient and cardiologist do not transmit a common secret key over the public channel. They use their indistinguishable valued weights as a common secret key which is essential for encryption and decryption. So, the likelihood of such attack to this proposed method is quite less with respect to other secret key exchange algorithms.

The future scope of our proposed technique is that TTNN model may be implemented in any online communication for the key distribution and certification purpose on any medical Expert System. In addition, this technique may be used as login password verification system in different E-Health Care. Also some of the evolutionary algorithm may be attached with this TTNN model to derive well uniform distribution of the weight vector.

Acknowledgements Our deep sense of gratitude to physionet database https://physionet.org/physiobank/database/stdb/. The signals from MIT-BIH ST Change Database had been used for experimental purpose.

References

1. Cryptography Key: Retrieved 06 Aug 2017, from http://en.wikipedia.org/wiki/Key_(cryptography)
2. Diffie, W., Hellman, M.: Multi-user cryptographic techniques. In: *Proceedings of the AFIPS Proceedings*, vol. 45, pp. 109–112, 8 June 1976

3. Diffie, W., Hellman, M.: New directions in cryptography. IEEE Trans. Inform. Theory **22**(6), 644–654 (1976)
4. Praveenkumar, P., Catherine Priya, P., Avila, J., et al.: Wireless Pers. Commun. (2017). https://doi.org/10.1007/s11277-017-4795-x. Springer US, Print ISSN 0929-6212, Online ISSN 1572-834X
5. Anusudha, K., Venkateswaran, N., Valarmathi, J.: Multimed. Tools Appl. **76**(2), 2911–2932 (2017). https://doi.org/10.1007/s11042-015-3213-1. Springer US, Print ISSN 1380-7501, Online ISSN 1573-7721
6. Hemeda, Afaf, Saif, Aasem, et al.: Simultaneous acute arterial and venous cerebral thrombosis and acute upper limb thrombotic ischemia due to combined uncommon hereditary factors for thrombophilia in young adult. J. Vascu. Med. Surg. **2017**, 297 (2017)
7. Seong, A.C., et al.: A review of coronary artery disease research in Malaysia. Med. J. Malays. **71**(Suppl), 42–57 (2016)
8. Nandavaram, S., Chandrasekar, V.T., Savici, D.: Acute pulmonary vascular talcosis: mimicking acute pulmonary embolism case report. J. Vasc. Med. Surg. **4**, 272 (2016)
9. Schummer, W.: Towards optimal central venous catheter tip position. J. Vasc. Med. Surg. **2016**, 260 (2016)
10. Al-Haj, A., Mohammad, A., Amer, A.: J Digit Imaging **30**(1), 26–38. https://doi.org/10.1007/s10278-016-9901-1
11. Sarkar, A., Mandal, J.K.: Computational science guided soft computing based cryptographic technique using ant colony intelligence for wireless communication (ACICT). Int. J. Comput. Sci. Appl. (IJCSA) **4**(5), 61–73 (2014). https://doi.org/10.5121/ijcsa.2014.4505. ISSN 2200-0011
12. Sarkar, A., Mandal, J.K.: Intelligent soft computing based cryptographic technique using chaos synchronization for wireless communication (CSCT). Int. J. Ambient Syst. Appl. (IJASA) **2**(3), 11–20 (2014). https://doi.org/10.5121/ijasa.2014.2302. ISSN 2321-6344
13. Sarkar, A., Mandal, J.K.: Secured transmission through multi layer perceptron in wireless communication (STMLP). Int. J. Mob. Netw. Commun. Telemat. (IJMNCT) **4**(4), 1–16. ISSN 1839-5678
14. Sarkar, A., Mandal, J.R.: Cryptanalysis of key exchange method using computational intelligence guided multilayer perceptron in wireless communication (CREMLP). Adv. Comput. Intell. Int. J. ACII **1**(1), 1–9 (2014). ISSN 2317-4113
15. Goldberger, A.L., Amaral, L.A.N., Glass, L., Hausdorff, J.M., Ivanov, P.C., Mark, R.G., Mietus, J.E., Moody, G.B., Peng, C.-K., Stanley, H.E.: PhysioBank, PhysioToolkit, and PhysioNet: components of a new research resource for complex physiologic signals. Circulation **101**(23):e215–e220. Circulation Electronic Pages; http://circ.ahajournals.org/content/101/23/e215.full, 13 June 2000

Part II
Circuit System and Devices

Computation of Peak Tunneling Current Density in Resonant Tunneling Diode Using Self-consistency Technique

Arpan Deyasi, Biswarup Karmakar, Rupali Lodh and Pradipta Biswas

1 Introduction

Design of low-power semiconductor devices is the need of the day for the last two decades, which gave birth of semiconductor heterostructure [1] with possible various potential applications following the remarkable work of Tsu and Esaki [2]. Supported by precisely controlled fabrication techniques [3], low-dimensional quantum devices already exhibited novel electronic [4] and optoelectronic [5] properties which are far superior to their bulk structures. In this context, computation of quantum transport process inside the device should be accurately performed as transmission coefficient and corresponding current density [6, 7] will determine its candidature for implementation purpose.

Growth of AlGaAs resonant tunneling diode by molecular beam epitaxy method is reported [8] a long days ago in the nanometric range, and negative conductance is observed. Later it is fabricated on Si epitaxial wafer [9] for MEMS sensor design. RTD oscillators are recently proposed as terahertz sources [10] for wireless communication applications. Photodetectors are designed using RTD at 1.3 μm [11] where sensitivity is calculated as a function of illumination power, and range is also calculated for constant sensitivity. Terahertz imaging system is constructed [12]

A. Deyasi (✉)
Department of Electronics and Communication Engineering,
RCC Institute of Information Technology, Kolkata, India
e-mail: deyasi_arpan@yahoo.co.in

B. Karmakar · R. Lodh · P. Biswas
Department of Electronic Science, A.P.C. College, Kolkata 700131, India
e-mail: biswarup.karmakar@yahoo.com

R. Lodh
e-mail: lodh.rupali.91@gmail.com

P. Biswas
e-mail: pradiptabiswas1994@gmail.com

© Springer Nature Singapore Pte Ltd. 2019
J. K. Mandal et al. (eds.), *Contemporary Advances in Innovative and Applicable Information Technology*, Advances in Intelligent Systems and Computing 812,
https://doi.org/10.1007/978-981-13-1540-4_4

with 6 dB enhancement of SNR. Very recently, super-harmonic GHz oscillations are observed [13] with optical feedback. Electroluminescence emission is observed [14] with higher order of optical on/off ratio in a wider temperature zone. Microwave-photonic interface and short-range high GB wireless links are developed [15] using RTD sources.

In the present paper, peak current density is computed for two different sets of RTD devices, and is plotted with structural parameters; temperature, material compositions, and applied bias are calculated. In Sect. 2, results are discussed in the light of significant differences that are observed for different dimensions of the devices. Results are crucial for its candidature in THz sources. In Sect. 3, summarization is made based on the analytical findings.

2 Results and Discussions

From Fig. 1 it is observed that maximum peaks are found for Al mole fraction 0.1, 0.2, 0.4 at 0.08 V. So, among three set of Al mole fraction maximum tunneling current density is achieved for Al mole fraction 0.2 which is 4.416×10^5 A m^{-2} at 0.08 V and the current densities are comparatively low for mole fraction 0.4 and 0.1. This is so because when any two eigenenergy state of DBQW structure matched with each other, the maximum transmission probability occurs, i.e., the quantum tunneling phenomenon is happened and maximum value of peak current

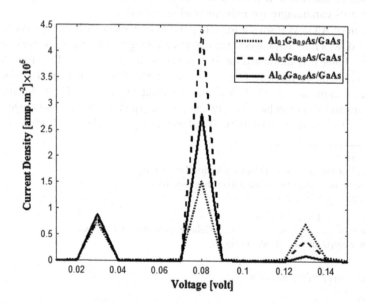

Fig. 1 Current density variation with applied voltage for different material compositions of barrier widths with Self-consistency technique (for $y = 0.2$)

density is achieved. Then further variation of voltage multiple peaks are achieved but their peak current densities are lower than maximum peak, because the carrier concentration is lowered for such energy state.

For L.B.W = 70 nm, R.B.W = 80 nm, the peak current density remain constant from mole fraction 0.5 to 0.1. After 0.1 the peak current density slowly increases up to 0.3, and then increases rapidly. For higher barrier dimension set quantum encirclement increases with increase the Al mole fraction from lower value to higher and the current density also increases.

Figure 2 shows the variation of peak tunneling current density with middle barrier width for two different set of material composition and same well width. Here exactly same curve is obtained for the two sets of mole fraction. The highest current density peak 10.66×10^4 A m^{-2} is achieved at 5 nm and further increase in barrier width, reduces the peak current density. Since with increasing barrier width dimension the quantum encirclement decreases and it does not remain confined to quantum region hence the current density decreases.

Figure 3 shows the variation of peak tunneling current density as a function of temperature for two different set of barrier width. Both the temperature variation curve for L.B.W = 70 nm, R.B.W = 80 nm and L.B.W = 30 nm, R.B.W = 40 nm are linear. In both cases, peak current density increases with increasing temperature. But the slope of the characteristics curve for L.B.W = 70 nm, R.B.W = 80 nm is less than that of the curve obtained for lower barrier dimension. This means that for L.B.W = 30 nm, R.B.W = 40 nm peak current density increases more with increasing temperature compare to the peak current variations for L.B.W = 70 nm,

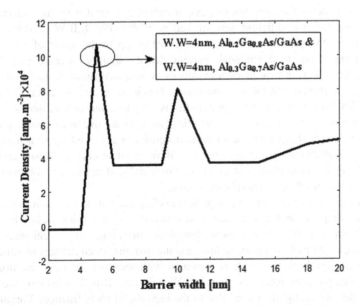

Fig. 2 Variation of Peak tunneling current density as a function of barrier width for two different set of material composition and same well width

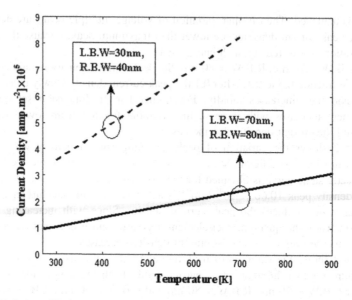

Fig. 3 Variation of Peak tunneling current density as a function of temperature for two different sets of barrier width

R.B.W = 80 nm. In this case for any set of barrier width the quantum confinement always increases with increase the temperature and current density in also better in high temperature.

Figure 4 shows the variation of peak tunneling current density as a function of well width for two different sets of barrier width. For L.B.W = 70 nm, R.B. W = 80 nm, the peak current density is constant up to the value of well width 4 nm. From 4 nm to 5 nm peak current density sharply increase and further increase of well width, peak current density remains constant up to 9 nm. After 9 nm peak current density sharply decreases. This is so because for a certain limit of well width dimension tunneling probability is very low and current density is also low After 4 nm the tunneling probability as well as quantum confinement is maximum and a sharp increase of current density is achieved up to a maximum value and further increase in well width it show constant current density and after 9 nm the quantum confinement decrease which reduced tunneling probability and hence the current density sharply decreases.

Figure 5 shows the variation of peak tunneling current density as a function of material composition for $x < y$ for two different set of barrier width. For L.B. W = 30 nm, R.B.W = 40 nm, peak tunneling current density is enhanced with increasing of Al mole fraction. In this case the quantum encirclement modified the transmission probability which is increases with increase the mole fraction and current density also increases. For L.B.W = 70 nm, R.B.W = 80 nm, the peak current density is slightly varied due to the increase of mole fraction. Variation of current is as smaller as it is considered as a constant variation. In this case quantum

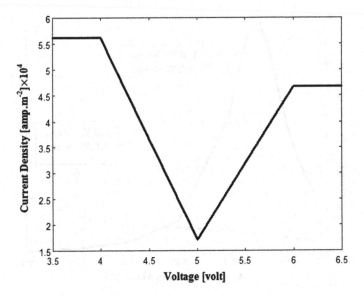

Fig. 4 Variation of Peak tunneling current density as a function of well width for the dipping section for L.B.W = 30 nm, R.B.W = 40 nm

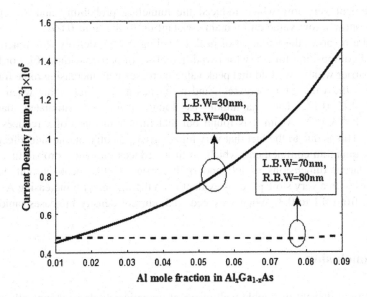

Fig. 5 Variation of Peak tunneling current density as a function of material composition ($x < y$) for two different set of barrier width

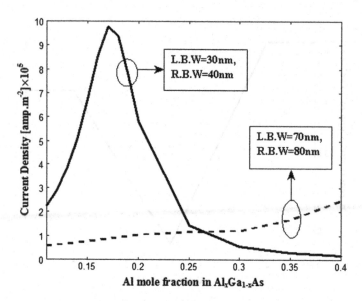

Fig. 6 Variation of Peak tunneling current density as a function of material composition ($x > y$) for two different set of barrier width

encirclement very low which reduced the tunneling probability and the current density show a low value and remain constant up to the limit 0.09.

Figure 6 shows the variation of peak tunneling current density as a function of material composition for $x > y$ for two different set of barrier width [16]. For lower set of barrier widths, we find that peak value increases with increasing mole fraction of Al (as the value of x increases), and it reaches a very high value when x lies within 0.15–0.17. The magnitude as found from the simulated data is 9.77×10^5 A m^{-2}. Again it is found out that further increase of x reduces peak density. This is due to the fact that very high barrier potential at one side effectively reduces quantum confinement, which, in turn, reduces tunneling probability.

For larger dimensions of left and right barrier widths, peak current density increases with a very slow pace, as evident from the figure, with increasing Al mole fraction from 0.1 to 0.3. When x exceeds 0.3, current density increases rapidly.

3   Conclusion

Peak current density of double well resonant tunneling diode is analytically investigated with different structural parameters, material compositions and also for other external effects. Different dimensional configurations are considered to study the effect in lower and moderate nanometric range. Significance of the present work lies in the fact that resonant diodes are operated at low biases when resonant tunneling is happened, and thus corresponding peak current flows through the quantum device at that condition. Hence the current plays the crucial role in determining the electronic behavior of the device. Thus, for any particular application, determination of magnitude of peak current becomes essential and its variation over a wider range of different interrelating functions is henceforth studied.

References

1. Talele, K., Patil, D.S.: Analysis of wavefunction, energy and transmission coefficients in GaN/AlGaN superlattice nanostructures. Prog. Electromagnet. Res. **81**, 237–252 (2008)
2. Esaki, L., Chang, L.L.: New transport phenomenon in semiconductor superlattice. Phys. Rev. Lett. **33**, 495–498 (1974)
3. Ahn, C.H., Kim, H.H., Park, S.H., Son, C.M., Go, J.S.: Fabrication and performance evaluation of thin film RTD temperature sensor array on a curved glass surface. J. Korean Soc. Vis. **9**, 34–39 (2011)
4. Scandella, L., Güntherodt, H.J.: Field emission resonances studied with dI/ds(V) and dI/dV (V) curves. Ultramicroscopy **42**, 546–552 (1992)
5. Jacobs, K.J.P., Stevens, B.J., Hogg, R.A.: Photoluminescence characterization of high current density resonant tunneling diodes for terahertz applications. IEICE Trans. Electron. **E99.C**(2), 181–188 (2016)
6. Christodoulides, D.N., Andreou, A.G., Joseph, R.I., Westgate, C.R.: Analytical calculation of the quantum mechanical transmission coefficient for a triangular, planar-doped potential barrier. Solid State Electron. **28**, 821–822 (1985)
7. Li, Z., Tang, H., Liu, H., Liang, Y., Li, Q., An, N., Zeng, J., Wang, W., Xiong, Y.Z.: Improving the peak current density of resonant tunneling diode based on InP substrate. J. Semicond. **38**(6), 064005 (2017)
8. Kikuchi, A., Bannai, R., Kishino, K.: AlGaN resonant tunneling diodes grown by RF-MBE. Phys. Status Solidi **188**(1), 187–190 (2001)
9. Li, J., Guo, H., Liu, J., Tang, J., Ni, H., Shi, Y., Xue, C., Niu, Z., Zhang, W., Li, M., Yu, Y.: GaAs-based resonant tunneling diode (RTD) epitaxy on Si for highly sensitive strain gauge applications. Nanoscale Res. Lett. **8**(1), 218 (2013)
10. Wang, J., Al-Khalidi, A., Alharbi, K.: High performance resonant tunneling diode oscillators as terahertz sources. In: 46th European Microwave Conference (2016)
11. Pfenning, A., Hartmann, F., Langer, F., Kamp, M., Höfling, S., Worschech, L.: Sensitivity of resonant tunneling diode photodetectors. Nanotechnology **27**(35), 355202 (2016)
12. Miyamoto, T., Yamaguchi, A., Mukai, T.: Terahertz imaging system with resonant tunneling diodes. Jpn. J. Appl. Phys. **55**(3), 032201 (2016)
13. DalBosco, A.K., Suzuki, S., Asada, M., Minamide, H.: Super-harmonic oscillations in a resonant tunneling diode with optical feedback. Nonlinear Opt. OSA Tech. Dig. NW4A.15 (2017)
14. Hartmann, F., Pfenning, A., SousaDias, M.R., Langer, F., Höfling, S., Kamp, M., Worschech, L., Castelano, L.K.: Temperature tuning from direct to inverted bistable electroluminescence in resonant tunneling diodes. J. Appl. Phys. **122**, 154502 (2017)
15. Wasige, E., Alharbi, K.H., Al-Khalidi, A., Wang, J., Khalid, A., Rodrigues, G.C., Figueiredo, J.: Resonant tunneling diode terahertz sources for broadband wireless communications. In: Proceedings of Terahertz, RF, Millimeter, and Submillimeter-Wave Technology and Applications X, vol. 10103, 101031J (2017)
16. Karmakar, B., Lodh, R., Biswas, P., Ghosal, S., Deyasi, A.: Computation of current density in double well resonant tunneling diode using self-consistency technique. In: International Conference on Modelling and Simulation, Nov 2017

Computing Surface Potential and Drain Current in Nanometric Double-Gate MOSFET Using Ortiz-Conde Model

Krishnendu Roy, Anal Roy Chowdhury, Arpan Deyasi
and Angsuman Sarkar

1 Introduction

In order to reduce the adverse effects due to scaling down the dimension of MOSFET, different multigate structures are already proposed [1–3], and it is reported that their electronic properties are improved than that obtained for single-gate device [4, 5]. Computation of drain current in this context plays a pivotal role for the comparative analysis, which is calculated by Pao-Sah integral [6] applied in the solution of Poisson's equation under appropriate boundary conditions. Though several literatures [7, 8] are available by considering the undoped DG MOSFET structure, but questions have been raised regarding the validity of the approximation, and hence lightly doped structures are now taken into account for surface potential and drain current computation. Henceforth, a few literatures are also published [9–11] where the light doping in the substrate is considered for electronic property estimation.

K. Roy (✉) · A. R. Chowdhury
Department of Electronic Science, A.P.C. College, Kolkata 700131, India
e-mail: krishnendu.physics94@gmail.com

A. R. Chowdhury
e-mail: analroychowdhury084@gmail.com

A. Deyasi
Department of Electronics and Communication Engineering,
RCC Institute of Information Technology, Kolkata, India
e-mail: deyasi_arpan@yahoo.co.in

A. Sarkar
Department of Electronics and Communication Engineering,
Kalyani Government Engineering College, Kalyani, India
e-mail: angsumansarkar@ieee.org

© Springer Nature Singapore Pte Ltd. 2019
J. K. Mandal et al. (eds.), *Contemporary Advances in Innovative and Applicable
Information Technology*, Advances in Intelligent Systems and Computing 812,
https://doi.org/10.1007/978-981-13-1540-4_5

Cobianu showed an analytical method for calculating surface potential of undoped DG MOSFET [12] and also emphasized the importance of substrate thickness. Ortiz-Conde later modeled the device for both symmetric and asymmetric cases [13] from Taur's charge based model. Later the effect of microelectronic fabrication process is modeled [14] in calculation of surface potential and transconductance. Different SOI structures [14, 15] are also considered for analytical modeling. Effect of high doping [16] a considered on SOI structures for surface potential. New algorithms are also proposed [17, 18] based on Newton–Raphson's method for surface potential calculation. For undoped structure, quasi-non-equilibrium condition is also considered [19] for both the type of carriers.

In the present paper, surface potential and drain current of lightly doped DG MOSFET is calculated using Ortiz-Conde model. Dielectric thickness and substrate thickness are varied to observe the effect, and corresponding pinch-off voltage is calculated. In Sect. 2, mathematical formulation is briefly described, results are analyzed in Sect. 3, and conclusion is given in Sect. 4.

2 Mathematical Formulation

We consider a lightly doped symmetric double-gate MOSFET with electric field along z-axis, and carriers are transported along y-axis. With gradual channel approximation, Poisson equation is given by

$$\frac{d^2\phi}{dz^2} = \frac{q}{\varepsilon} n_i \exp\left[\frac{\phi - \phi_c}{\phi_t}\right] \tag{1}$$

where φ_c is the quasi-Fermi potential for electrons inside the channel.

Potential balance equation for the structure is given by

$$V_{GS} - V_{fb} = \frac{\varepsilon}{C_{ox}} \frac{d\phi}{dz}\bigg|_{z=t/2} + \phi|_{z=t/2} \tag{2}$$

Under appropriate boundary condition, electric field and potential are respectively obtained as [9, 13]

$$\zeta = -\frac{d\phi}{dz} = -\sqrt{\frac{2k_B T n_i}{\varepsilon}}\left[\exp\left[\frac{\phi - \phi_c}{\phi_t}\right] - \exp\left[\frac{\phi_0 - \phi_c}{\phi_t}\right]\right] \tag{3}$$

$$\phi(z) = \phi_0 - 2\phi_t \ln\left[\cos\left(\exp\left[\frac{\phi_0 - \phi_c}{2\phi_t}\right]\sqrt{\frac{q^2 n_i}{2k_B T \varepsilon}}\right)z\right] \tag{4}$$

Introducing Lambert W-function, potential balance equation can be written as

$$\frac{V_{GS} - V_{fb}}{2\phi_t} = \frac{\phi_s}{2\phi_t} + \frac{\sqrt{2k_B T n_i \varepsilon}}{2\phi_t C_{ox}} \exp\left[\frac{\phi_s - \phi_c}{2\phi_t}\right] \sin\left(\sqrt{\frac{q^2 n_i}{2k_B T \varepsilon}} \exp\left[\frac{\phi_s - \phi_c}{2\phi_t}\right] z\right) \quad (5)$$

Rearranging, surface potential can be obtained in the following form [9, 13]

$$\phi_s = V_{GS} - V_{fb} - 2\phi_t W$$
$$\times \left[\frac{\sqrt{2k_B T n_i \varepsilon}}{2\phi_t C_{ox}} \exp\left[\frac{-\phi_c}{2\phi_t}\right] \sin\left(\sqrt{\frac{q^2 n_i}{2k_B T \varepsilon}} \exp\left[\frac{\phi_s - \phi_c}{2\phi_t}\right]\right) \exp\left[\frac{V_{GS} - V_{fb}}{2\phi_t}\right]\right]$$
$$(6)$$

Current can be computed from the knowledge on electric field as given by Eq. (3). Drain current for the device can be considered as due to both drift and diffusion components

$$I_{DS} = 2\mu_n \frac{W}{L} \int_0^{V_{DS}} \int_{\phi_0}^{\phi_s} \frac{qn}{\xi} d\phi dV \quad (7)$$

With mathematical rearrangement, finally we obtain [9, 13]

$$I_{DS} = \mu_n \frac{W}{L} \left[\begin{array}{l} 2C_{ox}\left[V_{GS}(\phi_{sD} - \phi_{sS}) - 0.5(\phi_{sD} - \phi_{sS})^2\right] \\ + \frac{4k_B T C_{ox}}{q}(\phi_{sD} - \phi_{sS}) + tk_B n_i \left(\exp\left(\frac{\phi_{0D} - V_{DS}}{\phi_t}\right) - \exp\left(\frac{\phi_{0S}}{\phi_t}\right)\right) \end{array}\right]$$
$$(8)$$

3   Results and Discussions

Using Eqs. (5) and (8), surface potential and drain current are computed as a function of gate-to-flatband voltage (V_{gf}) and drain-to-source voltage (V_{DS}) respectively. It may be noted in this context that effect of flatband voltage is taken into account to make the result more realistic.

In Fig. 1, it is seen that at lower values of V_{gf}, both surface potential and bulk potential remains same as the mobile charge accumulated at the surface is less than or compatible with the amount of charge at the bulk region. But as the magnitude of V_{gf} reaches near the threshold value, surface charge begins to dominate, as a consequence of which surface potential starts to increase. The gap is more pronounced at higher values of V_{gf}, since potential drop across the dielectric layer increases. The threshold voltage obtained from Fig. 1a is 0.47 V, which is very close to the previous obtained result of 0.48 V [20] calculated for undoped structure with identical structural configuration.

Figure 1a represents the surface potential for 1 nm oxide thickness, whereas Fig. 1b represents it for 5 nm width. It is observed from the comparative study between the plots that the gap between surface potential and bulk potential decreases at higher V_{gf}. This is due to the fact that increasing the dielectric thickness reduces the potential drop across the layer, and thus surface charge becomes comparable with the bulk charge.

Drain current is simulated and plotted in Fig. 2. Here I_D is plotted for three different oxide thicknesses with moderate to higher gate voltages. It is seen from the figure that with increasing gate voltage, pinch-off voltage shifts to the higher magnitude. Also with increasing oxide thickness, saturation current decreases as it is quite expectable due to the higher potential drop across the layer. It is seen that for $V_{GS} = 1$ V, I_{DS} becomes 20, 0.8 and 0.5 mA respectively for 1 nm, 3 nm and 5 nm oxide thickness. The important point may be noted in this context that pinch-off voltage increases with increasing oxide thickness. For $V_{GS} = 1$ V, pinch-off voltage is calculated as 0.4, 0.45 and 0.5 V for 1, 3 and 5 nm oxide thickness respectively. The result obtained [20] for undoped structure is 0.48 V which is in close agreement. Consequently, dynamic resistance computed for a particular gate voltage decreases with enhancement of dielectric thickness. From the plots, r_{ds} becomes 0.1, 0.025, 0.02 Ω respectively when $V_{GS} = 1$ V. Similar observation is made for other values of V_{GS}.

Next the effect of substrate thickness is estimated on surface potential and bulk potential, as revealed in Fig. 3. Results are also compared with Fig. 1a. In Fig. 1a, substrate thickness is considered as 10 nm, whereas in Fig. 3a, it is 12 nm, and in

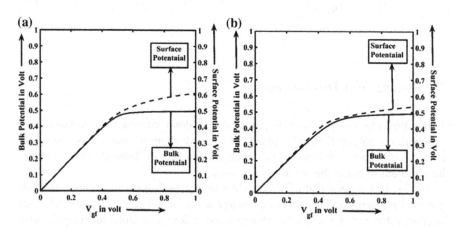

Fig. 1 Variation of surface potential w.r.t. bulk potential with V_{gf} for **a** 1 nm SiO$_2$ thickness; **b** 5 nm SiO$_2$ thickness

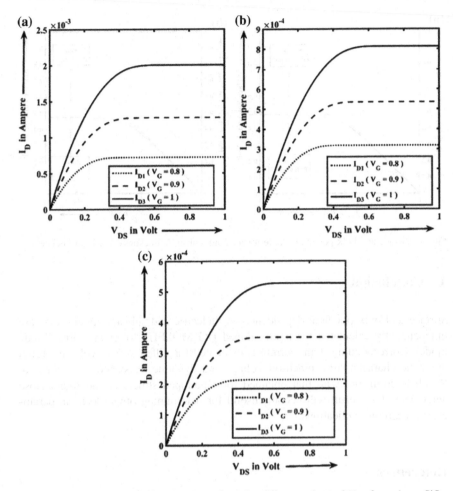

Fig. 2 Drain current with drain-to-source voltage for different values of V_{GS} for **a** 1 nm SiO_2 thickness; **b** 3 nm SiO_2 thickness; **c** 5 nm SiO_2 thickness

Fig. 3b, it is calculated as 15 nm. It is seen that both surface potential and bulk potential decrease with increasing substrate thickness. This is due to the fact that increasing substrate thickness reduces mobile charge density in the bulk region as the volume of bulk region increases. This causes lowering of bulk potential. Corresponding accumulation density at the surface decreases slowly. Result is shown below graphically.

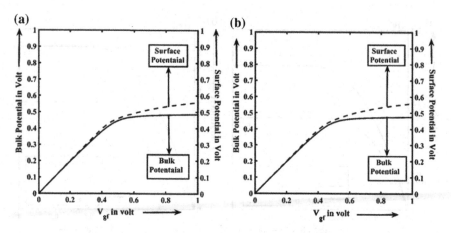

Fig. 3 Surface and bulk potential profile for **a** 12 nm substrate thickness; **b** 15 nm thickness

4 Conclusion

Surface and bulk potential dependence on dielectric and substrate thicknesses are independently calculated for lightly doped DG MOSFET following Ortiz-Conde model. Corresponding drain current is evaluated for different gate voltages. From the static characteristics, pinch-off voltage and dynamic resistance are computed. Result is compared with that obtained form undoped structure, and found close agreement. This result will become useful for determining other electrical parameters for circuit applications.

References

1. Taur, Y., Liang, X., Wang, W., Lu, H.: A continuous, analytic drain-current model for DG MOSFETs. IEEE Electron Device Lett. **25**(2), 107–109 (2004)
2. Subrahmanyam, B., Kumar, M.J.: Recessed source concept in nanoscale vertical surrounding gate (VSG) MOSFETs for controlling short-channel effects. Phys. E **41**(4), 671–676 (2009)
3. Jimenez, D., Iniguez, B., Sune, J., Marsal, L.F., Pallares, J., Roig, J., Flores, D.: Continuous analytic I–V model for surrounding-gate MOSFETs. IEEE Electron Device Lett. **25**(8), 571–573 (2004)
4. Li, Y., Chou, H.M.: A comparative study of electrical characteristic on sub-10-nm double-gate MOSFETs. IEEE Trans. Nanotechnol. **4**, 645–647 (2005)
5. Li, Y., Yu, S.M.: A unified quantum correction model for nanoscale single- and double-gate MOSFETs under inversion condition. Nanotechnology **15**, 1009–1016 (2004)
6. Pao, H.C., Sah, C.T.: Effect of diffusion current on characteristics of metal-oxide (insulator)-semiconductor transistors. Solid State Electron. **9**(10), 927–937 (1966)
7. Chang, S., Wang, G., Huang, Q., Wang, H.: Analytic model for undoped symmetric double-gate MOSFETs with small gate-oxide-thickness asymmetry. IEEE Trans. Electron Devices **56**(10), 2297–2301 (2009)

8. Yu, B., Lu, H., Liu, M., Taur, Y.: Explicit continuous models for double-gate and surrounding-gate MOSFETs. IEEE Trans. Electron Devices **54**(10), 2175–2722 (2007)
9. Ortiz-Conde, A., Garcia-Sanchez, F.J.: A rigorous classical solution for the drain current of doped double-gate MOSFETs. IEEE Trans. Electron Devices **59**(9), 2390–2395 (2012)
10. Lo, S.C., Li, Y., Yu, S.M.: Analytical solution of nonlinear Poisson equation for symmetric double-gate metal-oxide-semiconductor field effect transistors. Math. Comput. Model. **46**, 180–188 (2007)
11. Abebe, H., Cumberbatch, E., Morris, H., Tyree, V., Numata, T., Uno, S.: Symmetric and asymmetric double gate MOSFET modeling. J. Semicond. Technol. Sci. **9**(4), 225–232 (2009)
12. Cobianu, O., Glesner, M.: A computationally efficient method for analytical calculation of potentials in undoped symmetric DG SOI MOSFET. Techn. Proc. NSTI Nanotechnol. Conf. Trade Show Compact Model. **3**(7), 804–807 (2006)
13. Ortiz-Conde, A., Garcia-Sanchez, F.J., Muci, J., Malobabic, S., Liou, J.J.: A review of core compact models for undoped double-gates SOI MOSFETs. IEEE Trans. Electron Devices **54**(1), 131–140 (2007)
14. Nandi, A., Saxena, A.K., Dasgupta, S.: Analytical modeling of a double gate MOSFET considering source/drain lateral Gaussian doping profile. IEEE Trans. Electron Devices **60**(11), 3705–3709 (2013)
15. Sayed, S., Khan, M.Z.R.: Analytical modeling of surface accumulation behavior of fully depleted SOI 4 GATE transistors (G(4)-FETs). Solid State Electron. **81**, 105–112 (2013)
16. Jurczak, M., Jakubowski, A., Lukasiak, L.: The effects of high doping on the I-V characteristics of a thin-film SOI MOSFET. IEEE Trans. Electron Devices **45**(9), 1985–1992 (1998)
17. Jandhyala, S., Mahapatra, S.: An efficient robust algorithm for the surface-potential calculation of independent DG MOSFET. IEEE Trans. Electron Devices **58**(6), 1663–1671 (2011)
18. Zhou, X., Zhu, Z., Rustagi, S.C., See, G.H., Zhu, G., Lin, S., Wei, C., Lim, G.H.: Rigorous surface-potential solution for undoped symmetric double-gate MOSFETs considering both electron and holes at quasi nonequilibrium. IEEE Trans. Electron Devices **55**(2), 616–623 (2008)
19. He, J., Bian, W., Tao, Y., Liu, F., Lu, K., Wu, W., Wang, T., Chan, M.: An explicit current–voltage model for undoped double-gate MOSFETs based on accurate yet analytic approximation to the carrier concentration. Solid State Electron. **51**(1), 179–185 (2007)
20. Ortiz-Conde, A., Garcia-Sanchez, F.J., Muci, J.: Rigorous analytical solution for drain current of undoped symmetric dual-gate MOSFETs. Solid State Electron. **49**(4), 640–647 (2005)

Comparative Studies on the Performance Parameters of a P-Channel Tunnel Field Effect Transistor Using Different Channel Materials for Low-Power Digital Application

Jayabrata Goswami, Anuva Ganguly, Anirudhha Ghosal
and J. P. Banerjee

1 Introduction

MOSFETs can be scaled down to achieve high packing density and on-current. However the sub-threshold swing of MOSFETs cannot be scaled down below the thermionic limit of 60 mV/decade at room temperature. As a result, it becomes difficult to reduce the off-state leakage current below a certain limit. In this respect Tunnel Field Effect Transistors (TFETs) are very promising devices where band-to-band tunneling phenomenon is used to lower the sub-threshold swing below 60 mV/decade with high on–off current ratio and lower off-state leakage current [1, 2]. It is reported that TFETs with Si as channel material exhibit low on state current (100 $\mu A/\mu m$) [3]. If Ge having lower band gap than Si be used as channel material in TFETs, the on-state current increases to 850 $\mu A/\mu m$ [4]. Carbon-based material like graphene in the form of nanoribbon (GNR) can be used as 1-D channel in TFETs for better on-state current. The width tunable band gap of GNR helps to improve the ratio of on-current to off-current and sub-threshold swing for low-power performance of GNR TFETs in digital operation at lower supply voltages. Recent studies show that GNR having larger ribbon width possesses

J. Goswami (✉) · A. Ganguly · A. Ghosal · J. P. Banerjee
Institute of Radio Physics and Electronics, University of Calcutta, Kolkata, India
e-mail: goswamijayabrata@gmail.com

A. Ganguly
e-mail: gangulyanuva@gmail.com

A. Ghosal
e-mail: aghosal2008@gmail.com

J. P. Banerjee
e-mail: scope.jcb@gmail.com

© Springer Nature Singapore Pte Ltd. 2019
J. K. Mandal et al. (eds.), *Contemporary Advances in Innovative and Applicable Information Technology*, Advances in Intelligent Systems and Computing 812,
https://doi.org/10.1007/978-981-13-1540-4_6

smaller E_G and lower effective masses, which leads to an increase of on-current, I_{on} and also decreases the leakage current, I_{off} [5, 6]. The purpose of this paper is to present a comparative study of the performance of TFETs using GNR as channel material with those using Si, Ge, InAs and InSb as regards sub-threshold swing, ratio of on-current to off-current and gate capacitance.

2 Device Structure and Self-consistent Model for Numerical Simulation of GNR PTFET

The cross sectional view of GNR PTFET is shown in Fig. 1. The TFET has highly doped n$^+$ source, p$^+$ channel and p$^+$ drain regions. 1D Poisson equation for the device is numerically solved to obtain the energy band diagrams for both on- and off-states and surface potential. The channel is fully depleted in the off-state at zero gate voltage. In the present work, the oxide thickness (t_{ox}) is taken to be 2 nm. The gate oxide material of the device is Y_2O_3.

The channel is assumed to be fully depleted both in the off-state at zero gate to source voltage and also in the on-state with low gate to source voltage. Both the gate and drain are reverse biased with respect to source. So the reverse bias gate to source and drain to source voltages are V_{GS} and V_{DS} respectively. The p$^+$ channel is taken in positive x-direction whose length is L_{CH}.

The surface potential at a particular position of channel of GNR PTFET can be obtained from numerical solution of 1D Poisson's equation subject to the appropriate boundary conditions.

$$\frac{d^2\varphi_{surf}(x)}{dx^2} - \frac{\varphi_{surf}(x) - V_G - V_{BI}}{\lambda^2} = -\frac{q\rho(x)}{\varepsilon_{GNR}}, \tag{1}$$

where V_G is the gate potential, V_{BI} is the built-in potential, $\varphi_{surf}(x)$ is surface potential at position x, λ is the screening length for the particular device structure, $\rho(x)$ is the total charge density and ε_{GNR} is the permittivity of GNR. The total charge density is approximately equal to the impurity charge density since the source and drain regions are highly doped and channel is assumed to be fully depleted both at zero gate potential and also at low values of V_{GS}. Graphene

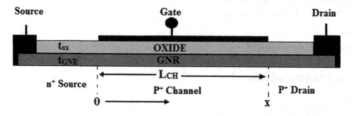

Fig. 1 Cross-sectional view of GNR PTFET

nanoribbon can be obtained by cutting two dimensional graphene sheet in narrow stripes which leads to lateral confinement, and induces a bandgap. The ribbon width dependence of band gap is explained below.

The energy dispersion relation for the GNR is given by [7]

$$E(n, k_x) = s\hbar v_F \sqrt{k_x^2 + k_n^2}, \tag{2}$$

where the value of s is $+1$ for the conduction band and -1 for the valence band, $\hbar$ is the reduced Planck's constant, v_F is the Fermi velocity of carrier in graphene, taken to be 10^6 m/s and k_n is the transverse wave vector corresponding to nth subband.

The transverse wave vector, k_n is given by

$$k_n = n\pi / 3w_{GNR}, \tag{3}$$

where $n = \pm 1, \pm 2, \pm 4, \pm 5, \pm 7,\dots$ and w_{GNR} is the ribbon width.

The electron momentum along the axis of GNR is $\hbar k_x$. Thus the band gap energy of GNR is related to ribbon width as follows:

$$\xi_G = \xi_C - \xi_V = \xi_{S=+1} - \xi_{S=-1} = \frac{2\pi\hbar v_F}{3w_{GNR}} \tag{4}$$

Here the width of GNR is much less than its length. Thus the bandgap of GNR depends on the ribbon width w_{GNR} and Fermi velocity (v_F).

The expression for screening length λ is given by [8]

$$\lambda = \sqrt{\left(\frac{\varepsilon_{GNR}}{\varepsilon_{ox}}\right) t_{GNR} t_{ox}} = \sqrt{t_{GNR} t_{ox}} \tag{5}$$

In the above expressions, t_{GNR} and t_{ox} are the thickness of GNR and gate oxide respectively. The permittivity of GNR and gate oxide are ε_{GNR} and ε_{ox} respectively. The relative permittivity of GNR is taken to be 16 [9] and that of high-k dielectric material (Y_2O_3) is 16 as used in the analysis. The high-k dielectric reduces the leakage current from gate to channel. The performance parameters of GNR PTFET such as ratio of on-current to off-current, sub-threshold swing or slope, intrinsic gate delay and gate capacitance are optimized by varying the channel length, ribbon width and gate oxide thickness of the device.

The energy band profile of GNR PTFET is obtained from numerical solution of quasi 1D Poisson's equation in source, channel and drain regions subject to the following boundary conditions:

(a) The electric field falls to zero at the highly doped source and drain ends of the channel.

(b) Both the electric field and potential are continuous at the source to channel and drain to channel interfaces.

(c) The separation of Fermi energy level from the top of the conduction band and the bottom of the valance band is taken to be equal to the thermal energy of the carrier.

(d) At zero gate bias the valence band in the channel is coincident with the Fermi level at the source end.

The tunneling current from the source to drain is calculated using Landauer formula [10] given by.

$$I_D = \frac{q}{\pi \hbar} \int (f_D(\xi) - f_S(\xi)) T_S(\xi) d\xi \tag{6}$$

Numerical integration of the product of tunneling probability, $T_S(\xi)$ and the difference between the F-D distribution function at the source and drain regions is carried out to obtain the drain current, I_D.

3 Simulation Results

Figure 2a, b show the simulated energy band diagrams for the optimized GNR PTFET in the on state (at $V_{GS} = -0.1$ V and $V_{DS} = -0.1$ V) and off-state ($V_{GS} = 0$ V) corresponding to the values of w_{GNR} and L_{CH} as 4 and 20 nm respectively. The Fermi energy levels (E_F) in the energy band plots are shown for both the off and on states of the device.

The interband tunneling in TFET depends on the position of the energy band in the channel region with respect to that in the source and drain regions. Figure 2a shows that the applied gate to source voltage pulls up the energy band of the source to channel in the on state which gives rise to tunneling current.

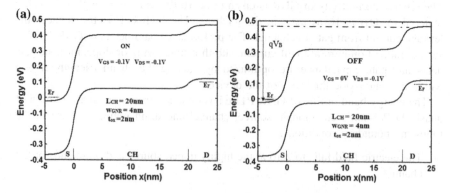

Fig. 2 Energy band plots for a GNR PTFET in the **a** on-state and **b** off-state

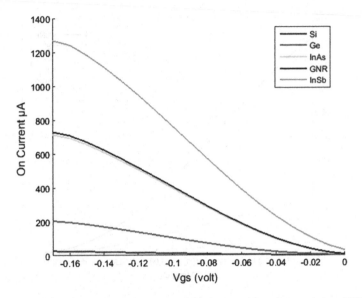

Fig. 3 The on-current or drain current versus gate to source voltage characteristics of PTFET for different channel materials

Figure 3 shows the plots of on-state current I_{on} against V_{gs} of the device for different channel materials.

The sub-threshold slope or swing, SS of a tunnel field effect transistor is related to the transconductance (g_m) and sub-threshold drain current (I_D) [11] as given by;

$$SS = \ln 10 \left(\frac{g_m}{I_D} \right) \tag{7}$$

Figure 4 shows the ratio of on-current to off-current versus gate to source voltage (V_{GS}) for different channel materials. The ratio of on-current to off-current is found to be higher in GNR TFET as compared to that in Si, Ge, InAs and InSb TFETs. At $V_{GS} = -0.1$ V, this ratio is found to be 2500, 850, 250, 245, and 49 using GNR, Si, Ge, InAs and InSb as channel materials respectively. Further the sub-threshold slope (SS) of the optimized PTFET with GNR as channel material is found to be 8 mV/decade at $V_{GS} = -0.1$ V which is lower than that with other channel materials and therefore GNR TFETs are very much suitable for low-power digital application.

The intrinsic gate delay (τ_{int}) is an important parameter defined as the ratio of the total induced channel charge to the on-state threshold current. The following expression is used to obtain (τ_{int}) [12].

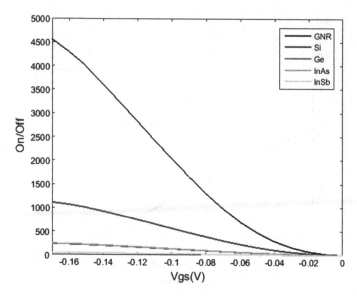

Fig. 4 Ratio of on-current to off-current for different channel materials

$$\tau_{int} = (Q_{C(on)} - Q_{C(off)})/I_{on}, \qquad (8)$$

where $Q_{C(on)}$ and $Q_{C(off)}$ are the on-state and off-state charges in the channel respectively while I_{on} is defined the on-state current. The supply voltage is equal to $(V_{on} - V_{off})$ and taken to be 0.1 V in the calculation.

The gate capacitance (C_G) is the rate of change of total channel charge (Q_C) with V_{GS} and given by

$$C_G = dQ_C/dV_{GS} \qquad (9)$$

The magnitudes of C_G for different values of gate-source voltages (V_{GS}) are calculated from the sum of gate-source and gate-drain capacitances, i.e., $C_G = C_{GS} + C_{GD}$.

Figure 5 and Table 1 show that the gate capacitance of PTFETs using GNR and InAs as channel materials are same i.e., 50 aF/μm while that using Si, Ge and InSb as channel material is higher, i.e., 59, 57 and 60 aF/μm, respectively.

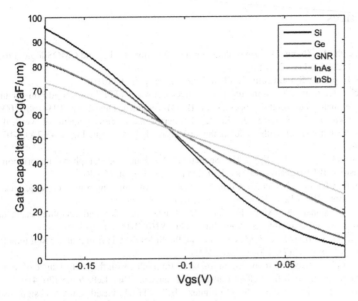

Fig. 5 Gate capacitance for different channel materials PTFETs

Table 1 At $V_{GS} = -0.1$ V and $V_{DS} = -0.1$ V

Channel material	τ_{int} (fs)	I_{ON}/I_{OFF}	Sub-threshold slope (SS)	Gate capacitance (C_G) (aF/μm)
InAs	2.0255	245	10 mV/decade	50
InSb	1.2570	49	12 mV/decade	60
GNR	1.9921	2500	8 mV/decade	50
Si	4.0309	850	NA	59
Ge	5.5271	250	20 mV/decade	57

4 Conclusion

A p$^+$ type Tunnel Field Effect Transistor using Si, Ge, InAs, InSb, and GNR as channel materials has been designed and modeled using MATLAB based software to study their comparative performance for low-power digital application. It is observed that PTFET with 1-D graphene nanoribbon as channel material provides best performance as regards on–off current ratio, low intrinsic gate delay, and low gate capacitance. It may therefore be concluded that GNR PTFET is the most promising one for low-power digital application.

References

1. Seabaugh, A., Zhang, Q.: Low-voltage tunnel transistors for beyond CMOS logic. Proc. IEEE **98**(12), 2095–2110 (2010)
2. Jeon, K., et al.: Symposium on VLSI Technology, p. 121 (2010)
3. Nirschl, T., et al.: The tunneling field—effect transistor (TFET) as an add-on for ultra-low voltage analog and digital processes. In: IEDM Technical Digest, pp. 195–198 (2004)
4. Bhuwalka, K.K., Schulze, J., Eisele, I.: Performance enhancement of vertical tunnel field-effect transistor with SiGe in the δp^+ layer. Jpn. J. Appl. Phys. **43**(7A), 4073–4078 (2004)
5. Zhang, Q., Fang, T., Xing, H., Seabaugh, A., Jena, D.: Graphene nanoribbon tunnel transistors. IEEE Electron Device Lett. **29**(12), 1344–1346 (2008)
6. Zhao, P., Chauhan, J., Guo, J.: Computational study of tunneling transistor based on graphene nanoribbon. Nano Lett. **9**(2), 684–688 (2009)
7. Fang, T., Konar, A., Xing, H., Jena, D.: Carrier statistics and quantum capacitance of graphene sheets and ribbons. Appl. Phys. Lett. **91**(092109), 1–3 (2007)
8. Yan, R.-H., Ourmazd, A., Lee, K.F.: Scaling the Si MOSFET: from bulk to SOI to bulk. IEEE Electron Device **39**(7), 1704–1710 (1992)
9. Kliros, G.S.: Analytical modeling of uniaxial strain effects on the performance of double-gate graphene nanoribbon field-effect transistors. Nanoscale Res. Lett. **9**, 65 (2014)
10. Barboni, L., Siniscalchi, M., Rodriguez, B.S.: TFET based circuit design using the transconductance generation efficiency g_m/I_d method. J. Electron Devices Soc. **3**(3) (2015)
11. Fahad, M.S., Srivastava, A., Sharma, A.K., Mayberry, C.: Analytical current transport modeling of graphene nanoribbon tunnel field-effect transistors for digital circuit design. IEEE Nanotechnol. **15**(1), 39–50 (2016)
12. Guo, J., Lundstrom, M., Datta, S.: Appl. Phys. Lett. **80**(17), 3192 (2002)

Implementation of Toffoli Gate Using LTEx Module of Quantum-Dot Cellular Automata

Chiradeep Mukherjee, Dip Ghosh, Sayan Halder,
Sambhu Nath Surai, Saradindu Panda, Asish Kumar Mukhopadhyay
and Bansibadan Maji

1 Introduction

Power dissipation is a key parameter in *"logical circuit design"* that attracts the researchers to design with the novel nanotechnology based Reversible Circuits. One-to-one mapping between the inputs and outputs in this computation technique enables the dissipation of heat energy within the SNL limits [1]. In recent years the work based on the Reversible Circuit synthesis using the Quantum Cellular Automata (QCA) has become emerging research thrust area in nanotechnology. QCA excels its superiority in terms of high packing density, extreme high speed and low power dissipation. In this technological approach, the information passes from input port to output port through the Columbic interaction between the electrons. Moreover the implementation of Reversible Logic by using QCA is under extensive practice [2] owing to the computational feasible nature of QCA. The Multiple Control Toffoli Gate (MCT) is most commonly used Gate in Reversible Computation where 3×3 Toffoli unit functions as the basic element [3]. The QCA is most promising and emerging technology to implement the Reversible circuits. This work focuses on the design of Area-Cost efficient 3×3 Toffoli Gate using LTEx Module [4] of QCA.

C. Mukherjee (✉) · B. Maji
Department of ECE, National Institute of Technology, Durgapur, India
e-mail: chiradeep.1234321@gmail.com

D. Ghosh · S. Halder · S. Panda
Department of ECE, Narula Institute of Technology, Kolkata, India

S. N. Surai · A. K. Mukhopadhyay
Kingstone Educational Institute, Kolkata, India

© Springer Nature Singapore Pte Ltd. 2019
J. K. Mandal et al. (eds.), *Contemporary Advances in Innovative and Applicable Information Technology*, Advances in Intelligent Systems and Computing 812,
https://doi.org/10.1007/978-981-13-1540-4_7

The entire work is organized as follows: Sect. 2 discusses about the background of QCA. The functionality of LTEx Module and its importance in QCA literature is presented in Sect. 3. The fundamentals of Toffoli Gate and existing QCA implementations of 3×3 Toffoli Gate are surveyed in Sect. 3.1. Following, the proposed LTEx 3×3 Toffoli Gate layouts and simulation results are presented in Sect. 3.2. The discussion about the design parameters of this work and implications are analyzed in Sect. 4. Finally Sect. 5 concludes the work and introduces some future extensions of the proposed 3×3 Toffoli Gate.

2 Background of QCA

The quantum cell which is the elemental square shaped structure of QCA comprises of four quantum dots surrounded by high potential barrier [5]. These dots are quantum-mechanically coupled such a way that an electron can tunnel only within the four quantum dots, not outside the square cell. A number of quantum cells form basic QCA Gates like (a) QCA Binary Wire, (b) Majority Voter and (c) Seven Cell Inverter as given in Fig. 1a–c respectively. Unlike conventional voltage or current flow, the information is likely to propagate from input to output through columbic interactions between the electrons of array of quantum cells. The QCA clock [6] helps the electrons to interact between themselves. Hence the orientations are given in Fig. 1a.

The Layered T Gate, in the field of QCA, has been introduced to be operated as universal NAND/NOR gate in QCA realization of Boolean Logic circuits [7]. The optimal realization of standard functions of QCA and n-bit Exclusive OR Gate [4] has been implemented using the Layered T Gate [7]. In reversible computation, the Toffoli Gate is used to realize the multilevel reversible circuits [8]. Since the Exclusive OR operation is the basic operation in reversible computation [9], the two-input Exclusive OR (Ex-OR) using Layered T Gate, namely LTEx Module [4] of Fig. 2, is extensively used to generate the QCA layout of 3×3 Toffoli Gate in this work. The LTEx module which has two inputs A0, A1 performs the Ex-OR operation and generates the output Z0 implementing Eq. (1):

$$Z0 = LTEx(A1, A0) = A1 \oplus A0 \tag{1}$$

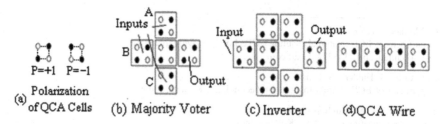

P=+1   P=−1

Polarization

(a) of QCA Cells   (b) Majority Voter   (c) Inverter   (d) QCA Wire

Fig. 1 QCA logic devices

Fig. 2 Two-input exclusive
OR using layered T Gate,
namely LTEx module [4]

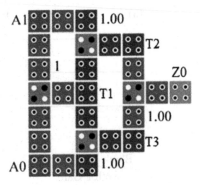

3 Realization of Toffoli Gate

The rudimentary problem of classical logic operation is the heat dissipation which is created by the information loss [10]. The reversible computation is the ultimate solution that avoids the objectionable features of the classical circuits. The one-to-one mapping between input-output and the absence of wire-loops are necessary facts during the implementation of multilevel reversible circuits. Like classical logic, the Reversible Benchmarks are introduced in Reversible circuit design paradigm to ease the complexity during reversible circuit realization. All the reversible library functions are generated using NCT (NOT, CNOT and Toffoli Gates), MCT (Multi Control Toffoli Gate), MCF (Multi Control Fredkin Gate) and EQ (Elementary Quantum Gate) reduction techniques [3]. The 3×3 Toffoli Gate is the basic concern of MCT reduction technique as well.

3.1 Toffoli Gate

The Toffoli Gate is introduced by T. Toffoli in the year 1982 [11]. The Toffoli Gate has three input vectors i.e. A2, A1, A0 and three output vectors i.e. Z2, Z1, Z0. The output Z0 inverts the input A0 into $\overline{A0}$ only when the other two inputs i.e. A2 and A1 become logic 1. The operation of Toffoli Gate is given in Eq. (2):

$$Z2 = A2, Z1 = A1 \text{ and } Z0 = A2 * A1 \oplus A0 \qquad (2)$$

The block diagram of 3×3 Toffoli Gate is shown in Fig. 3a. The input vector I(A2, A1, A0) and the output vector O(Z2, Z1, Z0) follow Eq. (2) to perform the operability of the Toffoli Gate as given in Fig. 3a. The QCA realization of 3×3 Toffoli Gate is shown in Fig. 3b.

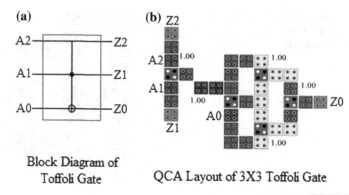

(a)

Block Diagram of
Toffoli Gate

(b)

QCA Layout of 3X3 Toffoli Gate

Fig. 3 a Block diagram of 3×3 Toffoli Gate, **b** QCA Layout of 3×3 Toffoli Gate

The QCA realization of 3×3 Toffoli Gate is implemented using two-input LTEx module of Fig. 2. The inputs i.e. A1 and A2 get multiplied to generate A2 * A1 by Layered T Inverter as indicated by part A. Next the product A2 * A1 is Ex-ORed with the input A0 and lastly the layout produce the output at Z0 as given by part B in Fig. 3b.

3.2 Comparison with the Existing Designs

The simulation waveform of proposed 3×3 Toffoli Gate which has been shown in Fig. 4 is generated with the Gallium Arsenide based hetero-structure parameters as mentioned in Table 1.

The QCA Layout of the proposed Toffoli Gate is compared with the existing 3×3 Toffoli designs [13–22] in terms of Effective area (in μm^2), O-Cost and Delay as reported in Table 2.

4 Result Discussion

The QCA Layout of 3×3 Toffoli Gate uses a single two-input LTEx Module followed by a Layered T Inverter. The proposed Toffoli design consumes 0.0388 μm^2 Effective Area that needs O-Cost of 36 and generates output after 0.75 clock cycle. The simulation output has been examined in bistable approximation based engine of QCA Designer as mentioned in Fig. 4. As an instance, the value of {111} at input vector is processed by the proposed Toffoli Gate to generate the value of {011} at output vector. Moreover, the proper analysis of proposed QCA

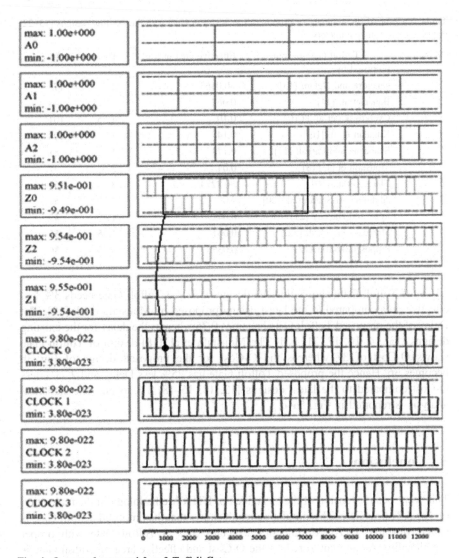

Fig. 4 Output of proposed 3 × 3 Toffoli Gate

Table 1 List of simulation parameters in QCADesigner [12]

Sl. No	Parameters	Value
1	Number of samples	12,800
2	Radius of effect	65 nm
3	Relative permittivity	12.9
4	Cell spacing	2 nm
5	Dot diameter	5 nm

Table 2 Comparison summary of proposed 3 × 3 Toffoli Gate with the existing designs [13–22] in terms of various parameters

Sl. No	3 × 3 Toffoli Gate design proposed by	Effective area in μm^2	Delay	O-Cost
1	Chandra and Netam [13]	>0.0563	1	174
2	Bahar et al. [14]	0.067	1	48
3	Cvetkovska et al. [15]	0.10	1.75	101
4	Kunalan et al. [16]	0.14	1	108
5	Shabeena and Pathak [17]	0.173	1	81
6	Moustafa et al. [18]	0.04	2	37
6	Chaves et al. [19]	>0.2274	3	155
7	Abdullah-Al-Shafi et al. [20]	0.06	3	57
8	Iqbal and Banday [21]	0.0427	1	45
9	Mahalakshmi et al. [22]	0.0614	1.25	55
10	Proposed Toffoli Gate	0.0388	0.75	36

layout of Toffoli Gate shows that the design of 3 × 3 Toffoli Gate needs 3% less effective area compared to the latest design proposed by Moustafa et al. [18]. The Delay of the proposed Toffoli Gate becomes 0.75 whereas the delay for the existing designs [13, 14, 16, 17, 21] is 1. The Moustafa et al. [18] design of Toffoli Gate shows O-Cost of 37 whereas the O-Cost for the proposed design of Fig. 3b becomes 36. Hence the proposed Toffoli Gate has 25% less Delay and 2.71% less O-Cost compared to the latest designs [13, 14, 16–18, 21] so far. The statistical analysis is reported in Fig. 5a–c.

5 Conclusion

This work is an extension of the basic idea of LTEx Methodology in the realization of 3 × 3 Toffoli Gate. The simulation result shows that the proposed Toffoli Gate is using two-input LTEx Module that executes 3 × 3 Toffoli Gate with proper functionality. An attempt to reduce the O-Cost and effective area of Toffoli Gate is made during the QCA layout design. This work can be further extended towards the design of generic Toffoli Gate to help in MCT realization of Reversible Circuits. The proposed Toffoli Gate could play the elementary role in the implementation of Reversible Benchmarks which will find extensive use in Reversible Digital Signal Processors.

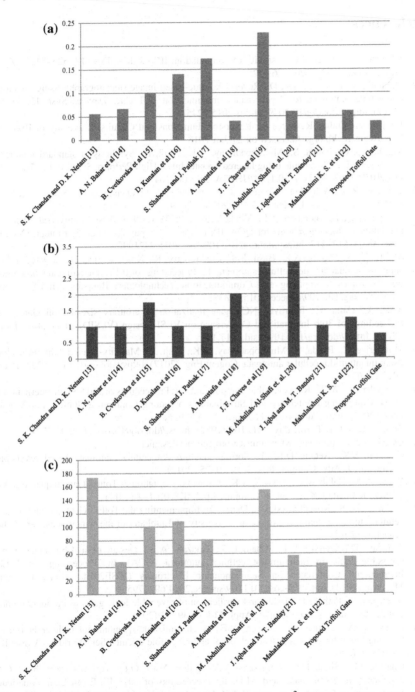

Fig. 5 a Analysis of effective area of proposed Toffoli Gate in μm^2 with other existing designs.
b Analysis of delay of proposed Toffoli Gate with other existing designs. **c** Analysis of O-Cost of
proposed Toffoli Gate with other existing designs

References

1. Bennett, C.H.: Logical reversibility of computation. IBM J. Res. Dev. **17**, 525–532 (1973). https://doi.org/10.1147/rd.176.0525
2. Debnath, B., Das, J.C., De, D.: Reversible logic-based image steganography using quantum dot cellular automata for secure nanocommunication. IET Circ. Devices Syst. **11**, 58–67 (2017). https://doi.org/10.1049/iet-cds.2015.0245
3. RevLib: An Online Resource for Reversible Functions and Circuits, University of Bremen. http://www.revlib.org. Accessed on 27 Jan 2018
4. Mukherjee, C., Panda, S., Mukhopadhyay, A.K., Maji, B.: Synthesis of standard functions and generic Ex-OR module using layered T Gate. Int. J. High Perform. Syst. Archit. **7**(2), 70–86 (2017). https://doi.org/10.1504/IJHPSA.2017.087164
5. Tougaw, P.D., Will, J.D., Graunke, C.R., Wheeler, D.I.: Quantum-dot cellular automata methods and devices. US Patent 7,602,207B2 (2009)
6. Campos, C.A.T., Marciano, A.L., Vilela Neto, O.P., Torres, F.S.: USE: a universal, scalable, and efficient clocking scheme for QCA. IEEE Trans. Comput. Aided Design Integr. Circuits Syst. **35**(3), 513–517. https://doi.org/10.1109/tcad.2015.2471996
7. Mukherjee, C., Sukla, A.S., Basu, S.S., Chakrabarty, R., Khan, A., De, D.: Layered T full adder using quantum-dot cellular automata. In: Proceedings of IEEE International Conference on Electronics, Computing and Communication Technologies, Bangalore, India (2015). https://doi.org/10.1109/conecct.2015.7383867
8. Kole, A., Datta, K.: Improved NCV gate realization of arbitrary size Toffoli Gates. In: Proceedings of 30th International Conference on VLSI Design (VLSID), Hyderabad, India (2017). https://doi.org/10.1109/vlsid.2017.11
9. Mukherjee, C., Panda, S., Mukhopadhyay, A.K., Maji, B.: Majority-layered T hybridization using quantum-dot cellular automata. Cogent Eng. (2017). https://doi.org/10.1080/23311916.2017.1286732
10. Drechsler, R., Wille, R.: Reversible computation. In: Proceedings of 2015 Sixth International Green Computing Conference and Sustainable Computing Conference (IGSC), Las Vegas, USA (2015). https://doi.org/10.1109/igcc.2015.7393687
11. Toffoli, T.: Int. J. Theor. Phys. **21**, 165 (1982). https://doi.org/10.1007/BF01857724
12. QCADesigner: [Online]. www.atips.ca/projects/qcadesigner
13. Chandra, S.K., Netam, D.K.: Exploring quantum dot cellular automata based reversible circuit. Int. J. Adv. Comput. Res. **2**(3), 70–75 (2012)
14. Bahar, A.N., Habib, MdA, Biswas, N.K.: A novel presentation of Toffoli Gate in quantum-dot cellular automata (QCA). Int. J. Comput. Appl. **82**(10), 1–4 (2013)
15. Cvetkovska, B., Kostadinovska, I., Danek, J.: Implementing the Toffoli Gate in quantum-dot cellular automata. Seminar project at University of Ljubljana in the winter semester of the academic (2013)
16. Kunalan, D., Cheong, C.L., Chau, C.F., Ghazali, A.B.: Design of a 4-bit adder using reversible logic in quantum-dot cellular automata (QCA). In: Proceedings of IEEE International Conference on Semiconductor Electronics (ICSE2014). Kuala Lumpur, Malaysia (2014). https://doi.org/10.1109/smelec.2014.6920795
17. Shabeena, S., Pathak, J.: Design and verification of reversible logic gates using quantum dot cellular automata. Int. J. Comput. Appl. **114**(4), 39–42 (2015)
18. Moustafa, A., Younes, A., Hassan, Y.F.: A customizable quantum-dot cellular automata building block for the synthesis of classical and reversible circuits. Sci. World J. (Article ID 705056) (2015)
19. Chaves, J.F., Silva, D.S., Camargos, V.V., Vilela Neto, O.P.: Towards reversible QCA computers: reversible gates and ALU. In: Proceedings of 2015 IEEE 6th Latin American Symposium on Circuits & Systems (LASCAS), Montevideo, Uruguay (2015). https://doi.org/10.1109/lascas.2015.7250458

20. Abdullah-Al-Shafi, M., Islam, M.S., Bahar, A.N.: A review on reversible logic gates and it's QCA implementation. Int. J. Comput. Appl. **128**, 27–34 (2015)
21. Iqbal, J., Banday, M.T.: Applications of Toffoli Gate for designing the classical gates using quantum dot cellular automata. Int. J. Recent Sci. Res. **6**(12), 7764–7769 (2015)
22. Mahalakshmi, K.S., Hajeri, S., Jayashree, H.V., Agrawal, V.K.: Performance estimation of conventional and reversible logic circuits using QCA implementation platform. In: Proceedings of 2016 International Conference on Circuit, Power and Computing Technologies (ICCPCT), Nagercoil, India (2016)

Study on Localized Surface Plasmon to Improve Photonic Extinction in Solar Cell

Partha Sarkar, Sambhu Nath Surai, S. Panda,
B. Maji and A. K. Mukhopadhyay

1 Introduction

With increasing global demand, we are first depleting the natural energy mostly fossil fuel, oil, gases, etc., and consequently becomes larger polluter and causes of ecological damages. Thus recent trends go towards green energy mainly renewable energies among of them solar energy is most promising. The incident solar energy hits on our earth surface are approximated 1650 TW per second which is more than combined with global energy consumption. Thus solar power will make the most of fulfillment of our increasing demand for large-scale production.

Photovoltaic solar cell converts phonic energy into electrical. In 1954, the demonstration of semiconductor-based solar cell is investigated with starting efficiency approx 6% in Bell Laboratories. After that to still now there are so many proposals has been coming for solar cell designing based on used materials, techniques to improve efficiency (η) etc. Today commercially available market leading c-si solar cell gives efficiency approx 14.7% by [1], using back surface field

P. Sarkar (✉) · B. Maji
Department of ECE, National Institute of Technology, Durgapur 713209, India
e-mail: parthasarkar.info@gmail.com

B. Maji
e-mail: bmajiecenit@yahoo.com

S. N. Surai · S. Panda
Department of ECE, Narula Institute of Technology, Kolkata 700109, India
e-mail: sambhu.surai@gmail.com

S. Panda
e-mail: saradindupanda@gmail.com

A. K. Mukhopadhyay
Kingston Educational Institute, Kolkata 700129, India
e-mail: askm55@gmail.com

© Springer Nature Singapore Pte Ltd. 2019
J. K. Mandal et al. (eds.), *Contemporary Advances in Innovative and Applicable Information Technology*, Advances in Intelligent Systems and Computing 812,
https://doi.org/10.1007/978-981-13-1540-4_8

(BSF) gives efficiency reaches to 15.5% as [2] and another technique like rear local contact (RLC) gives solar cell efficiency reaches to 20%.

Currently, our achievable solar cell efficiency is approx 16–18%. In the solar photovoltaic cell, if incident photonic energy is more than or equal to the band-gap energy, the solar energy is absorbed and generates an electron–hole pair. Otherwise, there occur sub-band losses. Plasmonic is a newest emerging design techniques that may be a possible way to improvement in photonic absorption in solar cell and enhance absorption by photonic scattering due to metallic nanoparticles which are excited at surface Plasmonic resonance [3]. So the presence of metallic nanoparticles, the magnitude of Raman scattering is increased. This scattering provides more photon will be excited surface plasmon and improving quantum efficiency in the solar cell.

2 Plasmonics

Plasmon is a collective oscillation of electron cloud which oscillates in sync. The plasmonic phenomenon occurs mainly in metals, where valence electrons are loosely bounded with parent atoms [4]. Plasmons are two classes—localized surfaces plasmon and propagating plasmon. The localized surface plasmon occurs due to the dipole and or multipole oscillation of electron which sync at locally at the metal-dielectric interface. A typically localized plasmon presence in the electric field is shown in Fig. 1 as in [5].

2.1 Surface and Particle Electron Oscillation

The plasmonic effect occurs mainly in metals where conduction electron are loosely bounded and these electron clouds are described with Jellium model. So according to classical mechanics, in the presence of the positive electric field, the conductive

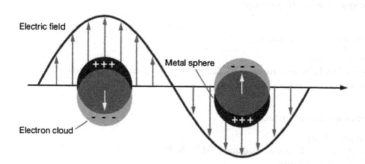

Fig. 1 Surface plasmon due to oscillation of electron in an electric field

electrons are shifted to x and in the absence of the field, the electrons are shifted back to its original position. By this way, the electrons will oscillate by the electric field due to the surface charges. Using classical theory applied to a single electron.

$$m\frac{d^2x}{dt^2} = -eE_x \tag{1}$$

From Gauss's theorem, it is considering that the electric field generated by a sheet carrying a surface charge nex will be $E_x = 2nex/2\varepsilon_o$. So, the collective oscillation of the electrons in the bulk and the plasma frequency are ω_P. In this study, we have considered metallic nanosphere, whose radius is very smaller than the incident wavelength. So the generated field due to uniform polarization Px of the particle $E_x = -Px/3\varepsilon_o$. So, the resonance frequency of the Plasmon in nanosphere is given by,

$$\omega_{sp} = \frac{\omega_p}{\sqrt{3}} = \sqrt{\frac{ne^2}{3m\varepsilon_o}} \tag{2}$$

2.2 Surface Plasmon

Surface plasmon polariton is a form of surface waves that propagate between metal–dielectric interface [6]. So for surface plasmon polariton, the choice of metals and dielectric are very important whose dielectric constant is described by Drude model to explain the transport properties of the electron and also this dielectric constant has strongly frequency dependence written as

$$\varepsilon_r(\omega) = 1 - \frac{\omega_p^2}{\omega^2 + i\gamma(\omega)\omega} \tag{3}$$

The localized surface modes that can propagate along a flat interface while decaying exponentially on both sides of the interface (Fig. 2).

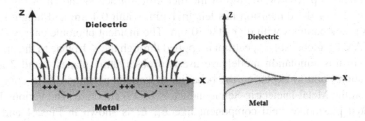

Fig. 2 Surface modes of propagation at the metal-dielectric interface

The extinction enhancement in metallic nanoparticles into a photovoltaic cell takes place due to photonic scattering and near-field concentration of light. The metallic nanoparticle is strong scatters with a photonic wavelength near the plasmon resonance [5]. From Mie theory, the cross sections (σ) of scattering and extinction are calculated as

$$\sigma_{\text{sca}} = \frac{\lambda^2}{2\pi} \sum_{n=0}^{\infty} (2n+1)\left(|a_n|^2 + |b_n|^2\right) \quad \sigma_{\text{ext}} = \frac{\lambda^2}{2\pi} \sum_{n=0}^{\infty} (2n+1)\text{Re}(a_n + b_n) \quad (4)$$

Here a_n and b_n are known as Mie coefficients and n is indexing running from 1 to ∞. Now, the extinction efficiency is written as $Q_{\text{ext}} = \sigma_{\text{ext}}/\pi r^2$, where '$r$' is the radius of nanosphere. If the diameter of the metallic nanoparticle is much smaller than the 1/10th wavelength of the incident photon, the cross-section scattering and absorption depend on polarizability (α) of the nanoparticles given by.

$$\alpha = 4\pi r^3 \frac{\varepsilon_{\text{sp}}(\omega) - \varepsilon_{\text{em}}}{\varepsilon_{\text{sp}}(\omega) + 2\varepsilon_{\text{em}}} \tag{5}$$

Here $\varepsilon_{\text{sp}}(\omega)$ is frequency dependent dielectric function of metallic nanoparticle and ε_{em} is a dielectric constant of embedding medium. If denominator term $\varepsilon_{\text{sp}}(\omega) + 2\varepsilon_{\text{em}}$ is minimum, resonant enhancement happens [6]. Now considering metallic nanoparticle with spherical in shape with volume $V = 4/3\pi r^3$ and dielectric function $\varepsilon_{\text{sp}}(\omega) = \varepsilon_R + i\varepsilon_I$. So the efficiency of extinction is given by.

$$Q_{\text{ext}} = \frac{72\pi r}{3\lambda} \varepsilon_{\text{em}}^{3/2} \frac{\varepsilon_I}{(\varepsilon_R + 2\varepsilon_{\text{em}})^2 + \varepsilon_I^2} \tag{6}$$

3 Plasmonic Solar Cell

In this study, we consider finite-difference time domain based solar cell model for studying plasmonic nanostructure for manipulating radiative and absorption properties of light trapping. In this model, we investigate the improvement photonic propagation into a thin crystalline-Si substrate where an array of the noble metal nanoparticle is placed on the top of the dielectric surface as shown in Fig. 3.

Here, the model dimension has length 1 μm, width 0.4 μm and height 0.6 μm with Ag nanoparticles with diameter 30 nm. The incident photonic energy equivalent to AM1.5 global solar spectrum is applied through the X-Y plane to the proposed model. In this simulation model, we use two observation planes 1 and 2 and an observation point to exploring the behavior of various plasmonic field component based on the Metal-Dielectric-Semiconductor (MDS) structure. The various FDTD simulated plasmonic field component like E_x, E_y is shown in Figs. 4 and 5 for

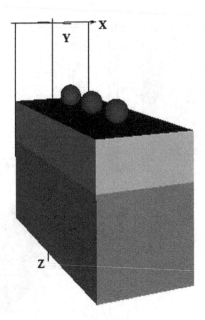

Fig. 3 FDTD model with Ag nanosphere radius 30 nm

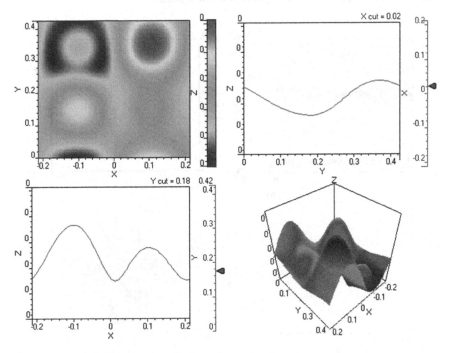

Fig. 4 Plasmonic field components E_x from observation plane 1

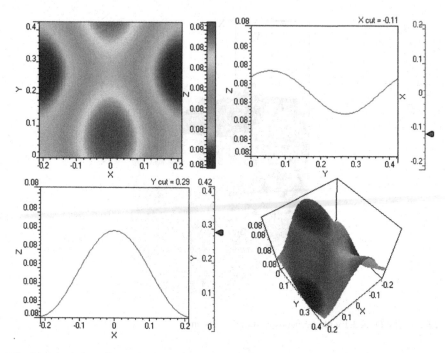

Fig. 5 Plasmonic field components E_y from observation plane 1

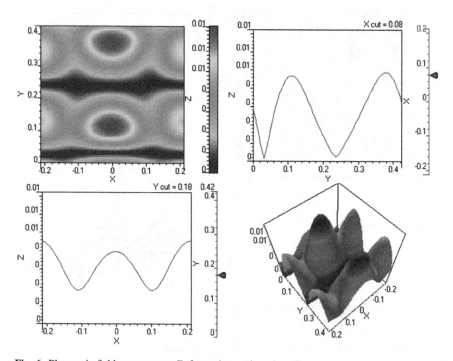

Fig. 6 Plasmonic field components E_z from observation plane 2

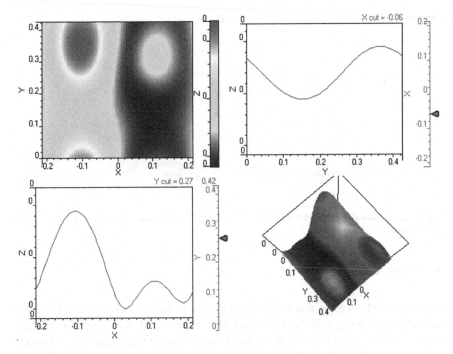

Fig. 7 Plasmonic field components H_y from observation plane 2

observation plane 1 and the field E_z, H_y is shown in Figs. 6 and 7 for observation plane 2.

Now to calculate the plasmonic resonance enhancement, we consider the FDTD simulated data for extinction efficiencies which are plotted for 10 nm to the 120 nm radius of noble metallic nanoparticle shown in Fig. 8.

The enhancement in extinction is calculated from the peak efficiency of a noble metal that has taken on the basis of optical absorption of bare Si as given in [7, 8]. For gold nanoparticle with radius 10 nm-120 nm, the enhancement in absorption is 7.9, 10.9, 16.9, 13.8, 8.3, 7.2, and 6.8% respectively. Similarly for silver nanosphere with radius 10–120 nm, the enhancement in absorption is 11.9, 16.4, 18.8, 13.5, 8.3, and 7.1% respectively. So, the increase in the size of nanosphere there will be a sharp decrease in efficiency. From Mie theory, the photonic scattering will higher than absorption if the dimension of the nanoparticle increases. We also see that when the radius of the nanoparticle is smaller than 50 nm, extinction efficiency will be much more than scattering.

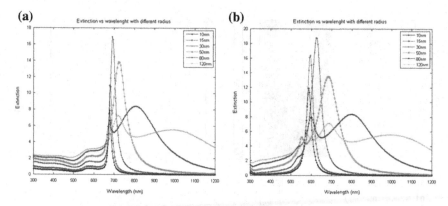

Fig. 8 Extinction efficiency versus incident wavelength for the various radius of **a** Au and **b** Ag nanoparticle

4 Conclusions

From our study, we see that when noble metal nanosphere radius near 30 nm; the extinction efficiency will be 16.89 and 18.78% for gold and silver nanoparticle respectively. It occurs due to absorption becomes greater than scattering. With the increase in radius, the photonic scattering will be dominated by extinction and hence decrease in efficiency. Here we see that both silver and gold nanosphere essentially improved the photocurrent and quantum efficiency of the thin-film solar cells. So, plasmonic becomes emerging technique to enhance photonic absorption with light trapping produced by scattering of incident light and improvement in efficiency.

References

1. Marrero, N., Gonzalez-Diaz, B., Lemus Guerrero, R., Borchert, D., Hernandez-Rodrıguez, C.: Optimization of sodium carbonate texturization on large-area crystalline silicon solar cells. Sol. Energy Mater. Sol. Cells **91**, 1943–1947 (2007)
2. Limmanee, A., Sugiura, T., Yamamoto, H., Sato, T., Miyajima, S., Yamada, A., Konagai, M.: Boron-doped microcrystalline silicon oxide film for use as back surface field in cast polycrystalline silicon solar cells. Jpn. J. Appl. Phys. **47** (2008)
3. Atwater, H.A.: The promise of plasmonics. Sci. Am. **56** (2007)
4. Enoch, S., Bonod, N.: Plasmonics From Basics to Advanced Topics. Springer, Berlin (2012)
5. Willets, K.A., Van Duyne, R.P.: Localized surface plasmon resonance spectroscopy and sensing. Annu. Rev. Phys. Chem. **58**, 267–297 (2007)
6. Maier, S.A.: Plasmonics: Fundamentals and Applications. Springer, New York (2007)
7. Palik, E.D.: Handbook of Optical Constants of Solids. Academic Press, Orlando (1985)
8. Johnson, P.B., Christy, R.W.: Optical constant of the Nobel metals. Phys. Rev. B. **6**(12) (1972)

Organic Electricity from Zn/Cu-PKL Electrochemical Cell

**K. A. Khan, M. S. Bhuyan, M. A. Mamun, M. Ibrahim,
Lovelu Hassan and M. A. Wadud**

1 Introduction

Bangladesh is one of the most populated as well as the least developed country. A report of Bangladesh Bureau of Statistics (BBS)-2005 suggested that around 40% of the total population live below the upper poverty line and about 25% people live below the lower poverty line [1]. Since Bangladesh is considered the most populated country with a huge population but only 49% people have entrance to grid electricity [2]. The people of Bangladesh who lives in off-grid areas but only around 25% of these people are using electricity [3–7]. With the growing concern of electrifying the rural areas, Bangladesh government has planned to find alternative sources of fuel rather than using gas and oil. Naturally, renewable energy is the most promising field and considered biomass as a part of it. Currently, Bangladesh is the seventh most crowded countries in this world and biomass provides 73% of

K. A. Khan (✉)
Department of Physics, Jagannath University, Dhaka 1100, Bangladesh
e-mail: kakhan01@yahoo.com

M. S. Bhuyan
Department of Business Studies, Limkokwing University of Creative Technology,
Cyberjaya, Malaysia
e-mail: awubsc@gmail.com

M. A. Mamun · M. Ibrahim
Department of Chemistry, Jagannath University, Dhaka, Bangladesh
e-mail: amamun@chem.jnu.ac.bd

L. Hassan
Department of Physics, Jahangirnagar University, Savar, Dhaka, Bangladesh
e-mail: lovelu.iu.bd@gmail.com

M. A. Wadud
Department of Chemistry, BAF Shaheen College, Dhaka, Bangladesh
e-mail: wadud_shimanta03@yahoo.com

© Springer Nature Singapore Pte Ltd. 2019
J. K. Mandal et al. (eds.), *Contemporary Advances in Innovative and Applicable Information Technology*, Advances in Intelligent Systems and Computing 812,
https://doi.org/10.1007/978-981-13-1540-4_9

the total energy [8–11]. Biomass promotes a clean and green environment as well as offers several advantages, e.g., high efficiency of energy conversion and the low cost of biomass residue. Biomass is already utilized in many countries and provides 50 EJ per year of total primary energy demand of this world [12, 13]. In Bangladesh, generation of electricity from biomass is very much available, environment friendly and easy to bearable cost [14–17]. A new type of renewable energy is invented in Bangladesh named, "Electricity generation from Pathor Kuchi Leaf (PKL)" [18–21]. It has been worked by the many researchers on PKL electricity, where the researchers have discussed regarding both their electrical and chemical properties for measuring, understanding, and developing the various parameters. In this paper, we have developed the correlations for producing voltage, current and power generation by introducing some equations theoretically and applying experimentally. The authors tried to find the use of secondary salt to enhance the power for large current. It was conducted the use of secondary salt and without secondary salt in the PKL extract for comparative study. It has been found that the findings of the research work will help the practical use of PKL electricity in the off-grid region. This work may play an important role for a new electricity generation technique in the world [22, 23].

2 Methodology

2.1 Specifications of PKL Juice Preparation

It was taken, PKL = 512 gm, H_2O = 610 gm, Waste/Residue = 82 gm. In order to produce electricity from the leaf of the *Bryophyllum*, first of all, its leaves have to be collected and then blended by blenders. Thereafter a mixture, containing pest and water with proportion generally 1:1, will have to be prepared. This mixture can be used directly for electricity production. This juice can be filtered out to get the clean juice for the use of electricity generation. Figure 1 illustrates the preparation process of juice. After blending the juice is poured and reserved in a plastic container or pot. This juice can be preserved for long-time use.

(a):PKLplant   (b):Extract machine   (c):PKL extract   (d):PKL Power output

Fig. 1 Methodology (**a–d**) of the PKL electricity generation system

2.2 Electromotive Force of the PKL Electrochemical Cell

The electromotive force (EMF) for the PKL electrochemical cell, E_{cell} can be written as [16]:

$$E^{o}_{cell} = E^{o}_{right\,(cathode)} - E^{o}_{left\,(anode)}$$

where, E^{o}_{cell} = Cell voltage at Standard state condition (at temperature 25 °C, pressure 1 atm and 1 M solution), which is called cell potential. $E^{o}_{right\,(cathode)}$ = Cathodic half-cell potential and $E^{o}_{left\,(anode)}$ = Anodic half-cell potential at Standard state condition. If $E^{o}_{right\,(cathode)} > E^{o}_{left\,(anode)}$, then E^{o}_{cell} = +ve and thus the reaction is spontaneous and the cell is referred to as Voltaic cell. If $E^{o}_{right\,(cathode)} < E^{o}_{left\,(anode)}$, then E^{o}_{cell} = −ve and hence the reaction is non-spontaneous and it is called electrolytic Cell. The PKL cell is constructed mainly based on Voltaic Cell [24, 25]. Since the PKl cell does not have any salt bridge, so it can be named as so-called Voltaic Cell or Quasi-Voltaic Cell.

2.3 Electrochemical Cells at a Glance

See Fig. 2.

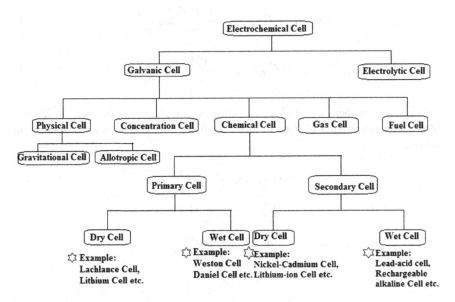

Fig. 2 Classification of electrochemical cells [26]

2.4 Electrochemistry

The redox reaction in this cell continues till the concentration of the ionic species (reactants and products) becomes equal to each other [5]. Thus the oxidation reaction takes place where the concentration of metal ions is low and the reduction takes place where the concentration is high.

The oxidation at anode:	$M - ne^- \rightarrow M^{n+}$
The reduction at cathode:	$M^{n+} + ne^- \rightarrow M$
The cell diagram:	$M/M^{n+} \mid M^{n+} \mid M$

2.5 PKL Cell Potential

We know, $E_{\text{Cell}} = E_{\text{right (cathode)}} - E_{\text{left (anode)}}$. So that,

$$E_{\text{Cell}} = E_{Cu^{2+}|Cu} + E_{2H^+|H_2} - 2E_{Zn^{2+}|Zn}, \tag{1}$$

where the symbols have their usual meanings. Now, we have,

$$E_{Cu^{2+}|Cu} = E^0_{Cu^{2+}|Cu} - \frac{0.0591}{n} \log \frac{1}{[Cu^{2+}]} \tag{2}$$

$$E_{2H^+|H_2} = E^0_{2H^+|H_2} - \frac{0.0591}{n} \log \frac{1}{[H^+]^2} \tag{3}$$

$$E_{Zn^{2+}|Zn} = E^0_{Zn^{2+}|Zn} - \frac{0.0591}{n} \log \frac{1}{[Zn^{2+}]^2} \tag{4}$$

where, n = number of moles of electrons transferred in cell reactions, $[Zn^{2+}]$ = Concentration of Zinc in molarity, $[Cu^{2+}]$ = Concentration of Copper in molarity,

$$E^0_{Zn^{2+}|Zn} = -0.76\,\text{V}$$

$$E^0_{Cu^{2+}|Cu} = 0.34\,\text{V}$$

$$E^0_{H^+|H_2} = 0.0\,\text{V}$$

$$So,\ PKL\ Cell\ potential = E^0_{Cu^{2+}|Cu} - \frac{0.0591}{n}\log\frac{1}{[Cu^{2+}]} - \left[E^0_{Zn^{2+}|Zn} - \frac{0.0591}{n}\log\frac{1}{[Zn^{2+}]}\right]$$

$$+ E^0_{2H^+|H_2} - \frac{0.0591}{n}\log\frac{1}{[H^+]^2} - - \left[E^0_{Zn^{2+}|Zn} - \frac{0.0591}{n}\log\frac{1}{[Zn^{2+}]}\right]$$

$$= E^0_{Cu^{2+}|Cu} + E^0_{2H^+|H_2} - 2E^0_{Zn^{2+}|Zn}$$

$$+ \frac{0.0591}{2}\left[\log\frac{1}{[Zn^{2+}]^2} - \left(\log\frac{1}{[Cu^{2+}]} + \log\frac{1}{[H^+]^2}\right)\right]$$

$$= 0.34 + 0 - 2(-0.76) + \frac{0.0591}{2}\log\frac{[Cu^{2+}][H^+]}{[Zn^{2+}]^2}$$

$$= 1.86 + \frac{0.0591}{2}\log\frac{[Cu^{2+}][H^+]}{[Zn^{2+}]}$$

$$(5)$$

Calculation of PKL Cell Potential:

Equation (5) represents the equation for the calculation of PKL cell potential. For a cell the cell potential is calculated below: where, the initial concentration of Zn^{2+} in molarity is = 0.0334 M, the initial concentration of Cu^{2+} in molarity = 0.3330 M, The pH of the solution = 3.5, the concentration of H^+ in molarity = antilog $(-3.5) = 3.16 \times 10^{-4}$ M, now putting the values in Eq. (5) we get,

$$PKL\ cell\ potential\ at\ temperature\ 25\,°C\ and\ 1\ atm = 1.86 + \frac{0.0591}{2}\log\frac{[0.333][3.16 \times 10^{-4}]}{[0.0334]}$$

$$= 1.85\,V.$$

Thus the theoretical value of PKL cell potential at temperature 25 °C and 1 atm is 1.85 for one unit cell. And thus the potential for a battery with six compartment = 6 × 1.85 V.

2.6 Determination of Equilibrium Constant of Cell Reaction

Equilibrium constant is the ratio of the equilibrium concentrations of the products raised to the power of their stoichiometric coefficients to the equilibrium concentrations of the reactants raised to the power of their stoichiometric coefficients.

For a reversible reaction, $aA + bB \leftrightarrow cC + dD$

$$K = [C]^c \cdot [D]^d / [A]^a \cdot [B]^b,$$

where,

[A] = equilibrium concentration of A in Molarity (M)
[B] = equilibrium concentration of B in Molarity (M)
[C] = equilibrium concentration of C in Molarity (M)
[D] = equilibrium concentration of D in Molarity (M)

And a, b, c, and d are the number of moles of A, B, C, and D respectively.

2.7   Theory for Determination of Equilibrium Constant

We know, from the thermo-dynamical principle,

$$W_{max} = -nFE_{max}, \tag{6}$$

where, n = Number of transferred electrons = 2. F = Faraday constant = 96,500 C, E = EMF of the PKL cell [17]. The output energy,

$$W = -nFE, \tag{7}$$

From the thermodynamics, we also know,

$$W_{max} = \Delta G, \tag{8}$$

where, W_{max} = Maximum Work, ΔG = Free energy for the chemical reaction in the PKL cell. Therefore, from (6) and (8) we can write,

$$\Delta G = -nFE_{max}, \tag{9}$$

again we have the Gibb's free energy,

$$\Delta G = -RT \ln K, \tag{10}$$

Therefore,

$$\Delta G = -2.303RT \log K. \tag{11}$$

Now, From (9) and (11) we get,

$\Delta G = -nFE_{max} = -2.303RT \log K$ or, $\log K = (-nFE_{max})/(-2.303RT \log K)$ or, $\log K = (nFE_{max})/(2.303RT \log K)$.

Therefore,

$$K = \text{antilog}(nFE_{max})/(2.303RT) \qquad (12)$$

2.7.1 Experimental Data for Calculation of Equilibrium Constant, K

Here, $F = 96,500$ C, $R =$ Molar Gas Constant $= 8.314$ JK^{-1} mol^{-1}, $n =$ Number of transferred electrons, which participated in the cell reactions. For cell-3 and 4 number of transferred electron, $n = 2$ and also $n = 4$ for other cells. $T =$ Room temperature $= 25$ °C $= 301$ K. Now the estimation of Equilibrium Constant (K): We have,

$$K = \text{antilog}(nFE_{max})/(2.303RT) \qquad (13)$$

$$E_{max} = \frac{E^0{}_{cell}}{6} V = X \text{ Volt (consider)},$$

where, $E^0{}_{cell}$ is the standard potential of the PKL module, Using these above values we can calculate the equilibrium constant. For a PKL cell ($E_{max} = 1.027$ V): $K = \text{antilog } (nFE_{max})/(2.303RT) = \text{antilog } \frac{4 \times 96500 \times 1.027}{2.303 \times 8.314 \times 301} = 6.03 \times 10^{68}$.

2.8 Comparative Study of the Discharge Characteristics of the Traditional Cells and PKL Cell Without Load

2.8.1 Discharge Characteristics of the Traditional Cells for Various Temperature Without Load for Comparative Study

In the absence of the external load, the capacity of a cell or battery can start decreasing. Because, Cell reaction happened inside the cell spontaneously.

Figure 3 shows the typical shelf life for some primary cells [24]: Typical self-discharge rates for common rechargeable cells are as follows. Nickel cadmium = 10%, Nickel Metal Hydride = 30% per month, Lithium = 5–10%, NiMH batteries = 1.25% per month.

Fig. 3 Variation of capacity with the variation of storage time

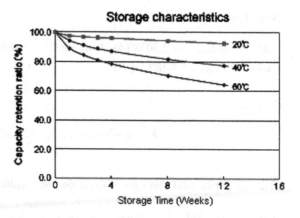

2.8.2 Self-Discharge Characteristics of PKL Cell for Comparative Study Without Load at Room Temperature

In this case, the electrolyte was prepared by the PKL extract mixing with water by filtration. The electrode was made by 1:1 Zn/Cu materials. PKL cell also loses its power with time for both Load and un-load conditions like other cells [25]. Though it is not quantify properly yet and at this stage it is not needed so much. But it shows a very interesting characteristic with time. If we keep it unused for some times its capacity regains [26]. Therefore, it shows better result on intermittent use with certain duration. As a result, it can be used for longer time instead of continuous use [27]. The experimental data regarding some performance parameters have been tabulated (Tables 1, 2, 3 and 4) and graphically (Figs. 3, 4, 5, 6, 7, 8, 9 and 10) discussed by the followings:

Table 1 Calculation of equilibrium constant

Cell No.	Temperature T (K)	Maximum potential of PKL battery E'_{max}	Maximum potential of PKL cell $E_{max} = E'_{max}/6$	Number of electron transferred, n	Equilibrium constant $K =$ antilog $(nFE_{max})/(2.303RT)$
1	301	5.89	0.982	4	5.88×10^{65}
2	,,	6.12	1.020	4	2.09×10^{68}
3	,,	4.93	0.822	2^a	3.39×10^{27}
4	,,	6.25	1.042	2^b	7.76×10^{34}
5	,,	6.34	1.057	4	6.16×10^{70}
6	,,	5.31	0.885	4	1.86×10^{59}
7		6.16	1.027	4	6.03×10^{68}

Where, [a]The PKL Cell was fuelled with only PKL juice and [b] the cell was fuelled with only secondary salt

Table 2 Determination of capacity of a PKL Cell

Time duration (h)	Current, I (A)	Capacity of the PKL cell (C = Ah)	Capacity rate (%) of the PKL cell
252	0.28	70.56	54.76
420	0.26	109.2	21.75
504	0.25	126	17.34
672	0.22	147.85	15.38
1000	0.18	180	13.77

Table 3 Determination of power of a PKL cell

Surface area of zinc (mm^2)	Short circuit current, I_{sc} (A)	Open circuit Voltage, V_{oc} (V)	P_{max} (W)
500	0.181	1.1	0.1991
600	0.186	1.098	0.2042
700	0.189	1.090	0.206
800	0.198	1.080	0.2138
900	0.206	1.070	0.2204
1000	0.209	1.066	0.2270
1100	0.211	1.060	0.2236
1200	0.218	1.058	0.2305 0.2306

Table 4 The decreasing of open circuit voltage (V) and short circuit current (A) with time (min) of PKL battery

Date	Local time	Total time (min)	Open circuit potential (V)	Short circuit current (A)
31.08.2014	10.36 am	00	6.03	5
	11.06 am	30	6.15	5
	11.36 am	60	6.07	5
	12.06 pm	90	6.09	5
	10.10 pm	694	5.47	1.5
	10.26 pm	710	5.47	1.3
01.09.2014	11.06 am	1470	5.36	0.5
	11.36 am	1500	5.35	0.5
	3.55 am	1759	5.31	0.4
	5.55 pm	1879	5.40	0.5
	6.55 pm	1939	5.32	0.5
	10.06 pm	2130	5.36	0.3
02.09.2014	8.20 am	2744	5.32	0.2
09.09.2014	6.18 pm	10,680	4.98	0.2

Fig. 4 Current (A) versus
time duration (h)

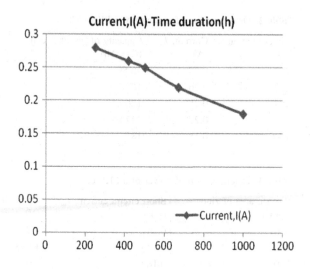

Fig. 5 Capacity (Ah) versus
time duration (h)

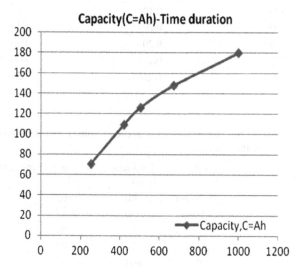

2.9   Determination of Self-discharging of PKL Cell with Secondary Salt

The specification of the electrolyte made by PKL extract with secondary salt is
given by the following composition: 40% PKL + 55% H_2O + 5% Secondary salt
($CuSO_4 \cdot 5H_2O$).

Fig. 6 Capacity rate (%)
versus time duration (h)

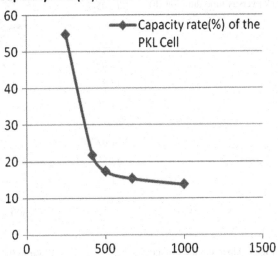

Fig. 7 Short circuit current
(A) versus time duration (h)

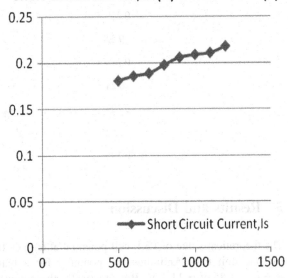

Fig. 8 Open circuit voltage
(V) versus time duration (h)

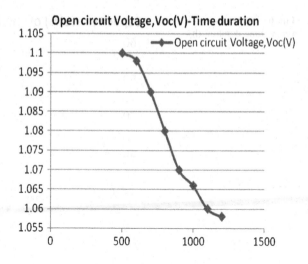

Fig. 9 Open circuit voltage
(V) versus time duration (h)

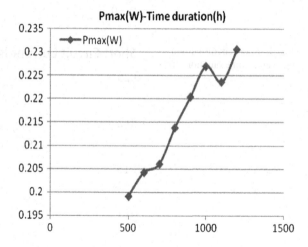

3   Results and Discussion

The theoretical value of PKL cell potential at temperature 25 °C and 1 atm is 1.85 for one unit cell. And thus the potential for a battery with six compartment is = 6 × 1.85 V = 11.1 V. But practically the potential is only 6.4 V. Thus it is seen that practically the cell potential does not depend on the acidity of PKL juice. Though the current flow can varies with the % of PKL juice. Where each of the electrodes was connected in parallel but when they were connected in series then we get the potential almost 11 V. Figure 4 shows the variation of Current with the variation of time. The current decreases slowly and it was almost constant at the beginning and slightly decreasing at the end. Figure 5 shows the variation of capacity with the variation of time for a PKL cell. The capacity increasers were

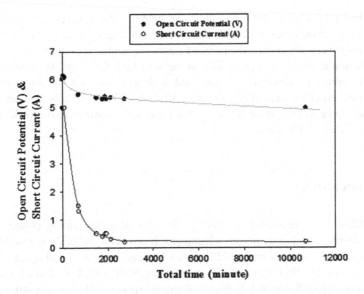

Fig. 10 Variation of open circuit voltage and short circuit current with time of the PKL battery

almost slight curvature at the beginning and was almost linear at the end. Figure 6 shows the variation of Capacity rate (%) with the variation of Time duration (h). It illustrates that the capacity rate (%) decreases rapidly at the beginning where as it was almost constant at the end. Figure 7 shows the variation of Short circuit current (A) with the variation of time duration (h). It was almost constant from the beginning to the end. Figure 8, the variation of Open circuit Voltage (V) with time duration (h), which shows that Open circuit Voltage (V) varies slightly that means almost constant. Figure 8 shows the variation of Power (W) without load with the variation of Time duration (h). It is also shown that the variation was almost constant during discharging period. Figure 10 shows the variation of Current and Voltage with the variation of time. PKL cell is a primary cell. So its life time to reserve the charge is considered to be high. But practically the life time is very much low [28]. From the data and figure it is demonstrated that both the potential and current is steady for few hours and after that both the parameter decrease sharply. However, the self-discharge rate is very high. This is due to the fact that both the electrodes are directly connected by fuelled solution (electrolyte) and so the reactive ions (H^+ and Cu^{2+}) undergo the reduction reaction at anode which may be named as local action. Thus due to unwanted chemical actions on the anode the PKL cell loses its energy. To prevent this unwanted energy losing and increasing the life time the electrodes may be separated by a separator. The reaction taking place is the ionic reaction where the charged species may produce and consume simultaneously. The concentration of product ions increases very fast and similarly the concentration of reactant ions decrease simultaneously. The equilibrium constant was calculated from the maximum potential which was observed at initial point. In Cell-3 and -4 there was only one type of ionic species demonstrated at Table 1 but for the other three cells there were both reactant species (H^+ and Cu^{2+}). From this we can decide that the forward reaction is higher for the reaction at which both the ionic species presence. This is because both Cu^{2+} and H^+ ions simultaneously reduce to give solid copper and hydrogen gas and for this zinc plate undergoes oxidation very much rapidly. Again $CuSO_4 \cdot 5H_2O$ acts as a secondary salt and hence the presence of this salt increases the ionization of weak organic acids presence in PKL juice.

4   Conclusions

This technology is feasible and viable for the off-grid regions in any country of the world. Because it has a numerous advantages. It is natural and clean renewable type with unlimited source of energy [29]. It can be generated again and again and will never run out. The PKL can cultivate everywhere even in the hilly and coastal areas of any country of the world. It does not require special skills for this simple PKL electric assembly; it is environment friendly and easy to operate this PKL System. Even a handicapped person can operate this PKL electric system. Last but not least, the PKL electricity generated by Zn/Cu-PKL (*Bryophyllum pinnatum* leaf) is cheaper than any other conventional portable battery and produces with LED lamp substantially cheaper lighting than kerosene [30]. This new technology may be the guide line for electricity production for the off-grid regions.

Acknowledgements The authors are grateful to the PKL electricity research group named A. K. M. Obaydullah, Dr. M. A. Latif, Dr. Md. Sajjad Hossain, Md. Mahfuz Alam, Dr. Fakhrul Islam, Bapy Guh, Md. Afzol Hossain, Dr. Jesmin Sultana and Prof. Dr. Mesbah Uddin Ahmed for their valuable suggestions and whole hearted cooperation during research work.

References

1. Khan, K.A., Alam, M.M.: Performance of PKL (Pathor Kuchi Leaf) electricity and its uses in Bangladesh. Int. J. Soc. Dev. Inf. Syst. **1**(1), 15–20 (2010)
2. Khan, K.A., Bakshi, M.H., Mahmud, A.A.: *Bryophyllum Pinnatum* leaf (BPL) is an eternal source of renewable electrical energy for future world. Am. J. Phys. Chem. **3**(5), 77–83 (2014) (Published online, 10 Nov 2014) (http://www.sciencepublishinggroup.com/j/ajpc). https://doi.org/10.11648/j.ajpc.20140305.15, ISSN: 2327-2430 (Print); ISSN: 2327-2449 (Online) (2014)
3. Paul, S., Khan, K.A., Islam, K.A., Islam, B., Reza, M.A.: Modeling of a biomass energy based (BPL) generating power plant and its features in comparison with other generating Plants . IPCBEE **44**(2012) @ (2012). IACSIT Press, Singapore https://doi.org/10.7763/ipcbee , V44. 3 (2012)
4. Khan, K.A., Paul, S., Zobayer, A., Hossain, S.S.: A study on solar photovoltaic conversion. Int. J. Sci. Eng. Res. **4**(3) (2013). ISSN2229-5518

5. Akter, T., Bhuiyan, M.H., Khan, K.A., Khan, M.H.: Impact of photo electrode thickness and annealing temperature on natural dye sensitized solar cell. J Elsevier. Ms. Ref. No.: SETA-D-16-00324R2 (2017)

6. Khan, K.A.: Inventors, electricity generation form Pathor Kuchi Leaf (PKL). Publication date 2008/12/31, Patent number BD 1004907

7. Khan, K.A.: Technical note Copper oxide coatings for use in a linear solar Fresnel reflecting concentrating collector, Publication date 1999/8/1. J. Renew. Energy **17**(4), 603–608 (Pergamon)

8. Khan, K.A.: Electricity generation form Pathor Kuchi Leaf (*Bryophyllum pinnatum*). Int. J. Sustain. Agril. Tech. **5**(4), 146–152 (2009)

9. Khan, K.A., Paul, S.: A analytical study on Electrochemistry for PKL (Pathor Kuchi Leaf) electricity generation system, Publication date 2013/5/21. In: 2013 IEEE Conference-Energytech, pp. 1–6 (IEEE)

10. Ruhane, T.A., Islam, M.T., Rahaman, M.S., Bhuiyan, M.M.H., Islam, J.M.M., Newaz, M.K., Khan, K.A., Khan, M.A.: Photo current enhancement of natural dye sensitized solar cell by optimizing dye extraction and its loading period. Optik Int. J. Light Electron. Opt. (2017) (Elsevier)

11. Khan, K.A., Alam, M.S., Mamun, M.A., Saime, M.A., Kamal, M.M.: Studies on electrochemistry for Pathor Kuchi leaf power system. J. Agric. Environ. **12**(1), 37–42 (2016) (Journal of Bangladesh)

12. Khan, K.A., Arafat, M.E.: Development of portable PKL (Pathor Kuchi Leaf) Lantern. Int. J. Soc. Dev. Inf. Syst. **1**(1), 15–20 (2010)

13. Khan, K.A., Bosu, R.: Performance study on PKL electricity for using DC Fan. Int. J. SOC. Dev. Inf. Syst. **1**(1), 27–30 (2010)

14. Khan, K.A., Hossain, M.I.: PKL electricity for switching on the television and radio. Int. J. Soc. Dev. Inf. Syst. **1**(1), 31–36 (2010)

15. Hasan, M., Hassan, L., Haque, S., Rahman, M., Khan, K.A.: A study to analyze the self-discharge characteristics of *Bryophyllum pinnatum* leaf fueled BPL test cell. J. IJRET **6** (12) (2017)

16. Sultana, J., Khan, K.A., Ahmed, M.U.: Electricity generation from Pathor Kuchi Leaf(PKL) (*Bryophyllum pinnatum*). J. Asiat Soc. Bangladesh Sci. **37**(4), 167–179 (2011)

17. Hasan, M., Haque, S., Khan, K.A.: An experimental study on the coulombic efficiency of *Bryophyllum pinnatum* leaf generated BPL cell. IJARIIE **2**(1) (2016). ISSN(O)-2395-4396

18. Khan, K.A., Paul, S., Adibullah, M., Alam, M.F., Sifat, S.M., Yousufe, M.R.: Performance analysis of BPL/PKL electricity module. Int. J. Sci. Eng. Res. **4**(3), 1 (2013). ISSN 2229-5518

19. Hasan, M.M., Khan, M.K.A., Khan, M.N.R., Islam, M.Z.: Sustainable electricity generation at the coastal areas and the Islands of Bangladesh using biomass resources. City Univ. J. **02** (01), 09–13 (2016)

20. Hasan, M., Khan, K.A.: *Bryophyllum pinnatum* leaf fueled cell: an alternate way of supplying electricity at the off-grid areas in Bangladesh. In: Proceedings of 4th International Conference on the Developments in Renewable Energy Technology [ICDRET 2016], p. 01 (2016). https://doi.org/10.1109/icdret.2016.7421522

21. Hasan, M., Khan, K.A., Mamun, M.A.: An estimation of the extractable electrical energy from *Bryophyllum pinnatum* leaf. Am. Int. J. Res. Sci. Technol. Eng. Math. (AIJRSTEM) **01** (19), 100–106 (2017)

22. Khan, K.A., Rahman, A., Rahman, M.S., Tahsin, A., Jubyer, K.M., Paul, S.: Performance analysis of electrical parameters of PKL electricity (An experimental analysis on discharge rates, capacity & discharge time, pulse performance and cycle life & deep discharge of Pathor Kuchi Leaf (PKL) electricity cell). In: 2016 IEEE Innovative Smart Grid Technologies-Asia (ISGT-Asia), pp. 540–544. IEEE (2016)

23. Khan, M.K.A., Paul, S., Rahman, M.S., Kundu, R.K., Hasan, M.M., Moniruzzaman, M., Al Mamun, M.: A study of performance analysis of PKL electricity generation parameters: (An experimental analysis on voltage regulation, capacity and energy efficiency of Pathor Kuchi

Leaf (PKL) electricity cell). In: 2016 IEEE 7th Power India International Conference (PIICON), pp. 1–6. IEEE (2016)

24. Hossain, M.A., Khan, M.K.A., Quayum, M.E.: Performance development of bio-voltaic cell from arum leaf extract electrolytes using zn/cu electrodes and investigation of their electrochemical performance. Int. J. Adv. Sci. Eng. Technol. **5**(4, Spl. Issue-1) (2017). ISSN: 2321-9009

25. Khan, K.A., Wadud, M.A., Obaydullah, A.K.M., Mamun, M.A.: PKL (*Bryophyllum pinnatum*) electricity for practical utilization. IJARIIE-ISSN(O)-2395-4396 **4**(1), 957–966

26. Khan, K.A., Rahman, A., Rahman, M.S., Tahsin, A., Jubyer, K.M., Paul, S.: Performance analysis of electrical parameters of PKL electricity (An experimental analysis on discharge rates, capacity & discharge time, pulse performance and cycle life & deep discharge of PathorKuchi Leaf (PKL) electricity cell). In: 2016 IEEE Innovative Smart Grid Technologies-Asia (ISGT-Asia), pp. 540–544. IEEE (2016)

27. Khan, M.K.A., Paul, S., Rahman, M.S., Kundu, R.K., Hasan, M.M., Moniruzzaman, M., Mamun, M.A.: A study of performance analysis of PKL electricity generation parameters: (An experimental analysis on voltage regulation, capacity and energy efficiency of Pathorkuchi leaf (PKL) electricity cell). In: 2016 IEEE 7th Power India International Conference (PIICON), pp. 1–6. IEEE (2016)

28. Khan, M.K.A., Rahman, M.S., Das, T., Ahmed, M.N., Saha, K.N., Paul, S.: Investigation on parameters performance of Zn/Cu electrodes of PKL, AVL, tomato and lemon juice based electrochemical cells: A comparative study. In: 2015 3rd International Conference on Electrical Information and Communication Technology (EICT), pp. 1–6. IEEE (2017)

29. Khan, K.A., Hassan, L., Obaydullah, A.K.M., Azharul Islam, S.M., Mamun, M.A., Akter, T., Hasan, M., Alam, M.S., Ibrahim, M., Rahman, M.M., Shahjahan, M.: Bioelectricity: a new approach to provide the electrical power from vegetative and fruits at off-grid region. J. Microsyst. Technol. Manuscript number: MITE-D-17-00623R2 (2018) https://doi.org/10.1007/s00542-018-3808-3 (Springer)

30. Khan, M.K.A, Rahman, M.S., Ahmed, M.N., Saha, K.N., Paul, S.: Investigation on parameters performance of Zn/Cu electrodes of PKL, AVL, tomato and lemon juice based electrochemical cells: A comparative study. J. Electr. Inf. Commun. Technol. (EICT). In: 2017 3rd International Conference on IEEE Xplore: 01 Feb 2018, https://doi.org/10.1109/eict.2017.8275150,ieee, Khulna, Bangladesh, Bangladesh, 7–9 Dec 2017

Part III
Nature Inspired Computing

Investigation of Dataset from Diabetic Retinopathy Through Discernibility-Based k-NN Algorithm

Rajesh Prasad Sarkar and Ananjan Maiti

1 Introduction

Definitely, one of vital organ around human body is eyes, through which numerous disease might be determined. The examinations of retina show prime sign of many diseases. Any injury to retina could be examined with the technique which is named Retinopathy. It primarily denotes retinal vascular disease followed by damage to the retina caused by abnormal blood flow. Among these, DR is the primary root of sightlessness within humans, which creates more than 5% of sightlessness according to the World Health Organization [1]. Diabetes can easily cause harm in the sights of the human beings with diseases which are called diabetic retinopathy. It is seen that excessive higher blood sugar level could cause impairment to blood vessels in the retina. The leak of blood vessels can also be seen. It is as well found that fresh blood vessels could develop on the retina. All of these kinds of deviations can decrease vision. This disease can impact individuals that are possessing dependency on genetic reasons, household activity, diet problem, and ecological issues. Indeed there can be two kinds of the reasons for type 1 as well as type 2 diabetes. The diabetes people mainly deals with difficulties sights of the eyes, head, heart, renal systems, feet, as well as nerves. The best approach to avoid or perhaps delay these complications is to assess at an early stage as well as have supervision of blood sugar level. This analysis requires computer-assisted examination process which would certainly predict and describe the various specifications of the

R. P. Sarkar (✉) · A. Maiti
Research Scholar of University of Engineering & Management (UEM), Kolkata, India
e-mail: rajesh.mca.cu@gmail.com

A. Maiti
e-mail: ananjan.maiti@gmail.com

A. Maiti
Department of IT, Techno India College of Technology, Kolkata, India

© Springer Nature Singapore Pte Ltd. 2019
J. K. Mandal et al. (eds.), *Contemporary Advances in Innovative and Applicable Information Technology*, Advances in Intelligent Systems and Computing 812,
https://doi.org/10.1007/978-981-13-1540-4_10

diseases. The role of the system is mere to take care of multiple attributes of the patient and identify the phase of diseases. This assessment will aid patient's confidence in the illness. Presently, machine learning system is employed in lots of applications that are associated with different fields which have the distinct attribute. Machine learning system refers to attaching tags to items that are characterized by a collection of good labeled features or characteristics. Indeed, there are three significant forms of pattern recognition trends. These are unlabeled classification such as unsupervised learning, labeled classification as supervised learning and in a mixture of both semi-supervised. To describe supervised learning techniques, we could focus on classification and regression method where each material of the information comprises a prior session label through the help of trainer or expert who has an idea about the formations of the classes. Here job of this instructor is to mention the correct answer to a classifier to carry out the labeling. This method combines the instructor's explication to generalize the issue and acquire its knowledge to learn the rules. This specific process could not be represented in a typical human frame and even it is complicated like Artificial Neural Networks (ANN) classifiers. The information of the instructor's labeling in the machine is to operate and understand the process rather than details of various functionality. In this way, the machine is learned the classification insights and enhanced the correctness of the classifier. The recommended approach reviewed the dataset of DR inmates with an altered k-NN algorithm. It comprises analysis of discernibility index (DI), adjustment of the dataset among this DI and as well implementation of a k-NN algorithm with the aid of new metric.

The other parts of the paper are framed as follows. The next section discusses some of the very current important literature regarding recognition of DR utilizing numerous machine learning methods figuring out their most encouraging concepts and also their disadvantages. Section 3 also presents different phases of methodology when it comes to assessing the classifiers on Dataset of Diabetic Retinopathy. The outcomes of observations carried out on these classifiers, as effectively as on k-NN, are shown in Sect. 4. Lastly, in Sect. 5 we have established our research with uncommon verdicts.

2 Related Works

Drall [2] focused on numerous types of eye diseases where means of fundus images deliberate several stages of Diabetic Retinopathy. This computer-based approach incorporates features extraction from these images. These characteristics are further assessed through customized SVM (Support Vector Machine). The research emphasizes the category of Normal, Non-proliferative diabetic retinopathy (NPDR) or proliferative diabetic retinopathy (PDR) affected eye with the precision of 94%. So, SVM with Consecutive minimal optimization algorithm results sufficiently as a classifier. In future, it should be analysed on more features and numerous varieties of devices.

Somasundaram et al. [3] thought about well known retinal vascular disorder such as DR as it triggers the blindness. The obtained fundus pictures reviewed via evaluating texture differentiation ability in which FI is distinguished along with the healthy images. The barrier was the higher dimensionality of the attributes. Their novel technique Machine Learning Bagging Ensemble Classifier (ML-BEC) is taken into consideration as finest suitable. The characteristics of DR disease evaluation includes optic nerve, blood vessels, neural tissue, as well as other associated information. The features extraction technique moved onward t-distributed Stochastic Neighbor Embedding (t-SNE). This Machine Learning techniques t-SNE uncovers a possibility of distribution across high-dimensional images along with dissimilar pairs. The automated testing system with Bagging Ensemble Classifier (BEC) assistance the base classifier which reveals much better classification accuracy (CA).

Saleh et al. [4] talked about the diagnosis of DR, which will certainly minimize the work and also conserve the amount of time of the medical professionals and the patients. They researched on procedures of managing various classes and this particular merged method improvisated the analysis outcome of two machine learning algorithms. The process selects the characteristic by breaking the tree-like structure. The benefit of a fuzzy method is which the choice is clear and also explainable to medical professionals and patients. In the near future, improvement can be feasible whenever they wanted to spread this particular merging method to the fuzzy random forest with fuzzy entropy and uncertainty of classification.

Carrera et al. [5] looked into upon DR that enhances the interest of the patients as a consequence of the loss of sight. The aim was actually to instantly categorize the training class of non-proliferative diabetic retinopathy through the retinal image. The characteristics of fundus images can be studied through a support vector machine to figure out the retinopathy group. This specific proposed strategy evaluated on a 400 retinal images of the database. The result of the assessment of non-proliferative diabetic retinopathy produced a sensitivity of 95% and a predictive capacity of 94%.

Lunscher et al. [6] observed that absence of professional specialists inside some mobile testing centers where enhancing DR testing is required. This operation could be much better accomplished with a computer-aided application on smartphones. Here these experts experimented on the function of SqueezeNet-based deep network accomplished on a fundus image dataset of near 88,000 for examination of diabetic retinopathy. The effects of this neural network validated the usefulness of mobile screening method of Diabetic retinopathy.

Adekunle et al. [7] proposed an intelligent methodology when it comes to medical diagnosis of diabetic retinopathy. It could be categorized into two classes non-proliferative diabetic retinopathy and proliferative diabetic retinopathy. The desired dataset applied for the use of this intelligent system was collected from UCI machine learning repository. The proposed system is going to assist medical professionals in diagnosing the illness effectively. This innovative design uses a feedforward neural network trained with backpropagation neural network. The results acquired in this task are compared to formerly proposed systems using the identical dataset Experimental outcomes suggests that our innovative system surpasses the various other methods of detecting diabetic retinopathy.

Ardiyanto et al. [8] conducted a research study through evaluating several features of the retinal images. This research study includes affordable a deep learning-based technique along with the embedded system which is an alternative to the medical professional when it comes to classification of the DR. This specific a cutting-edge methodology with a deep learning-based DR valuation for classifying its severity.

Shirbahadurkar et al. [9] cultivated of an automated scheme which is going to supports ophthalmologists in order to identify DR through several phases of the illness. This investigation designed a three-phase system in order to detect disease in the retinal fundus image. Initially, stage feasible morphological operations and Gabor filter used to filter out applicants. Soon after that, Feature vector using analytical, gray level and wavelet features for every candidate is designed. Inside the final phase, an association of these kinds of applicants as MAs and non-MAs is achieved using a multi-layered feed forward neural network (FFNN) classifier and support vector machine (SVM) classifier. The assessments have been carried out on the database DIARETDB1 to examine the suggested system. The assessment parameters accuracy, sensitivity, and specificity are obtained as 92, 79, 90%, and 95, 76, 92% respectively with regard to FFNN and SVM classifiers.

Abbasi-Sureshjani et al. [10] made use of convolutional neural networks which have enhanced the efficiency of image processing procedures significantly. Within this research, these experts explore the dataset with the deep residual networks to determine diabetes, without any having relevant information of glucose blood. The assessment reveals that convolutional networks are capable of grabbing attention as its efficiency of the proposed technique is really significantly higher than human specialists.

Mansour [11] established a CAD-based diabetic retinopathy (DR) which incorporates deep neural networks (DNNs). Within this study, AlexNet DNN that performs regarding the basis of the convolutional neural network (CNN) has been related to making it possible for an optimum DR CAD solution. The DR model uses a multilevel optimization method that includes preprocessing, adaptive-learning-based Gaussian mixture model (GMM)-based concept region segmentation, linked component-analysis-based area of interest (ROI) localization, AlexNet DNN-based highly dimensional feature extraction, principal component analysis (PCA)- and linear discriminant analysis (LDA)-based feature selection, and support-vector-machine-based classification to make sure optimal five-class DR distinction.

3 Methodology

In order to date set is needed which is directly obtained from UCI machine learning repository [12]. The following flowchart (Fig. 1) describes the proposed method of the classification. In this method, the k-NN classifier is used as a primary classifier. k-NN is a type of lazy learning algorithms which does not require any offline

Fig. 1 Proposed
methodology

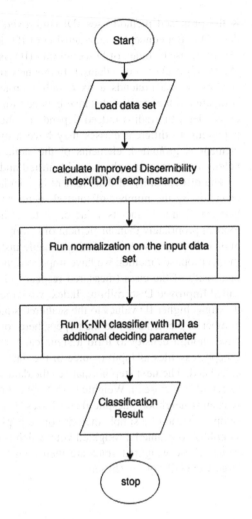

training. k-NN simply stores or memorizes all the training set at the time of training and during the classification stage for a particular testing instance it directly searches through all the training set and calculates the distances between the testing example and all the training set. These distances are then used to identify the nearest neighbors to classify the test instance. There are various methods to measure the distance, but among them, Euclidean and Minkowskian are the most popular. Finally to identify the class k-NN uses the K-nearest neighbors to vote among them to decide the class of the instance. Though k-NN has some advantages still this concept has a few inherent problems which are why many researchers have proposed different version and extension of the concept and tried to make the process more accurate and efficient. The major disadvantage of k-NN is its low accuracy as it takes only K as a decision making parameter. To improve the accuracy along with

K the proposed method uses IDI (Improved Discernibility index) to classify the data. This IDI concept is an extension of ID (Index of Discernibility (ID)) [13–15]. The basic use of Index of Discernibility (ID) is to measure the degree of distinguish ability of the classes of a dataset. In this new modification, we obtain use of (hyper) spheres; which pretends a fixed radius around each element of the dataset and compares to the average distance between this and the rest of the elements of that class. Here, the radius indeed depends on the formation of the class, so elements belonging to different classes may have a distinct radius. Once the radius of an element is grounded, elements of the same class as the examined element that belong to its (hyper) sphere are identified and counted.

The discernibility of that element is calculated by dividing the number of these elements by the number of total elements in the (hyper) sphere. So ID is index between 0 and 1, and its value close to 1 shows the element represents its class more appropriately than an element of lower value. In other words, an element with higher ID presents a more distinguishable instance from instances of other classes. In this proposed method we have improved the quality of ID by adding the concept of density of the class (measured respect to hypersphere) with ID. The new ID is called Improved Discernibility Index, which rectifies one disadvantage of ID where it assigns higher ID values to the scattered small group(small respect to the average number of elements inside the hypersphere) of same class elements which does not have enough density to form a significant cluster of class. Our proposed method accepts. csv files as input (source of the data), then the IDI of the whole dataset is calculated. The next step normalizes the data set, and in the last step, k-NN is used to classify. We used a Weighted k-NN where IDI acts as a weight. The usage of the K-nearest neighbors, despite their classes, yet then uses weighted votes from each sample instead of a simple majority or perhaps plurality voting rule. Every of the K samplings is granted a weighted vote which is equal to the value corresponding IDI value. These weighted votes are then total for each class, and the class with the largest overall vote is chosen.

4 Results

In this section, we estimate the performance of the computation of k-NN and proposed k-NN. The following actions were conducted on a Pentium Atom workstation with 2 GB main memory operating on Windows. All algorithms were performed in Python 3.6. In this experiment, we have worked on varying rates of k, and it is comprehended that proposed methodology works better in 5, 6, 7, values of k than original k-NN algorithm. Figs. 2 and 3 show different accuracy of the algorithm concerning 'k' value.

Fig. 2 This figure shows proposed k-NN achieves 72% accuracy

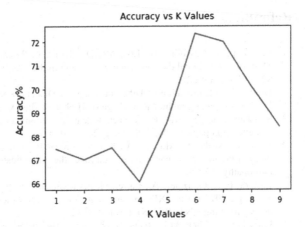

Fig. 3 This figure shows initial k-NN achieves 61% accuracy

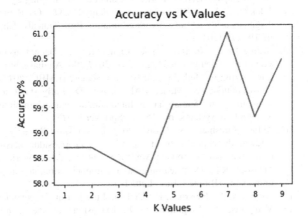

5 Conclusion

In this specific study, an inspection of DR employed via weighted k-NN classifier has been examined with specific preprocessing techniques including discernibility. Through estimation, instead of original k-NN, weighted k-NN classifier is optimal classifier when it comes to extracting and classifying the deviations in retina dataset. This provides 72% accuracy in the same dataset in which original k-NN gave 61%. This classifier accomplishes a higher efficiency since in review of DR which develops the diagnosis and testing of retinal diseases for the eye doctors in a more uncomplicated manner. In future it needs to be evaluated with other concerning classifier and modification could be possible in further.

References

1. Resnikoff, S., Pascolini, D., Etya'Ale, D., Kocur, I., Pararajasegaram, R., Pokharel, G.P., Mariotti, S.P.: Global data on visual impairment in the year 2002. Bull. World Health Organ. **82**, 844–851 (2002)
2. Drall, S.S.: Identification of Different Stages of Diabetic Retinopathy using Support Vector Machine. Chanderprabhu Jain College of Higher Studies & School of Law
3. Somasundaram, S., Alli, P.: A machine learning ensemble classifier for early prediction of diabetic retinopathy. J. Med. Syst. **41**, 201 (2017)
4. Saleh, E., Valls, A., Moreno, A., Romero-Aroca, P., Virgili, S.P.: Integration of different fuzzy rule-induction methods to improve the classification of patients with diabetic retinopathy
5. Carrera, E.V., González, A., Carrera, R.: Automated detection of diabetic retinopathy using SVM. In: 2017 IEEE XXIV International Conference on Electronics, Electrical Engineering and Computing (INTERCON), pp. 1–4. IEEE (2017)
6. Lunscher, N., Chen, M.L., Jiang, N., Zelek, J.: Automated screening for diabetic retinopathy using compact deep networks. J. Comput. Vis. Imaging Syst. **3** (2017)
7. Adekunle, A.A., Khashman, A., Olaniyi, E.O., Oyedotun, O.K.: Diabetic retinopathy diagnosis using neural network arbitration. Bull. Transilv. Univ. Brasov Math. Inf. Phys. Ser. III **10**, 179 (2017)
8. Ardiyanto, I., Nugroho. H.A., Buana, R.L.B.: Deep learning-based Diabetic Retinopathy assessment on embedded system. In: 2017 39th Annual International Conference of the IEEE Engineering in Medicine and Biology Society (EMBC, 2017), IEEE, pp. 1760–1763
9. Shirbahadurkar, S., Mane, V.M., Jadhav, D.: Early stage detection of diabetic retinopathy using an optimal feature set. In: International Symposium on Signal Processing and Intelligent Recognition Systems, pp. 15–23. Springer (2017)
10. Abbasi-Sureshjani, S., Dashtbozorg, B., ter Haar Romeny, B.M., Fleuret, F.: Exploratory study on direct prediction of diabetes using deep residual networks. In: European Congress on Computational Methods in Applied Sciences and Engineering, pp. 797–802. Springer (2017)
11. Mansour, R.F.: Deep-learning-based automatic computer-aided diagnosis system for diabetic retinopathy. Biomed. Eng. Lett. 1–17
12. Asuncion, A., Newman, D.: UCI machine learning repository (2007)
13. Voulgaris, Z., Magoulas, G.D.: Extensions of the k nearest neighbour methods for classification problems. In: Proceedings of the 26th IASTED International Conference on Artificial Intelligence and Applications (AIA), Innsbruck, Austria, 11 Feb 2008. pp. 23–28
14. Ishii, N., Morioka, Y., Bao, Y., Tanaka, H.: Control of variables in reducts-kNN classification with confidence. In: International Conference on Knowledge-Based and Intelligent Information and Engineering Systems, pp. 98–107. Springer (2011)
15. Voulgaris, Z., Magoulas, G.D.: A discernibility-based approach to feature selection for microarray data. In: 4th International IEEE Conference on Intelligent Systems, 2008, IS'08, pp. 21–22. IEEE (2008)

Malignant Melanoma Detection Using Multi Layer Perceptron with Optimized Network Parameter Selection by PSO

Soumen Mukherjee, Arunabha Adhikari and Madhusudan Roy

1 Introduction

Using Image processing tools some features of the morphology and texture of the skin become detectable which are not visible otherwise by naked eyes. In last decade quite a lot of researches have been done in the field of malignant melanoma identification using computer vision and machine learning technique as it is quick and can be done without the active involvement of the dermatologist with appreciably high accuracy. Malignant melanoma is increasing at a very fast rate of 3–7% throughout the world [1]. Melanoma can be cured if identified in early years which unfortunately very difficult due to inaccuracy in differentiating [2] from visually similar skin lesions like basal cell carcinoma, intraepithelial carcinoma, actinic keratosis, and squamous cell carcinoma which are of malignant type and seborrhoeic keratosis, melanocytic nevus, dermatofibroma, pyogenic granuloma, and haemangioma of benign type [3]. For diagnosis of malignant melanoma dermatologists need to relate many of their visual observations such as size, shape, color, border, and symmetry of the lesion area. Some popular scoring techniques used by the dermatologists for identifying malignant melanoma are 7-point checklist,

S. Mukherjee
Department of Computer Application, RCC Institute
of Information Technology, Kolkata, India
e-mail: soumou601@gmail.com

A. Adhikari (✉)
Department of Physics, West Bengal State University,
Barasat, West Bengal, India
e-mail: arunabha.adhikari@gmail.com

M. Roy
Surface Physics and Material Science Division,
Saha Institute of Nuclear Physics, Kolkata, India
e-mail: roy.madhusudan57@gmail.com

© Springer Nature Singapore Pte Ltd. 2019
J. K. Mandal et al. (eds.), *Contemporary Advances in Innovative and Applicable
Information Technology*, Advances in Intelligent Systems and Computing 812,
https://doi.org/10.1007/978-981-13-1540-4_11

ABCD rule, Menzies method, 3-point checklist [4]. Out of which ABCD rule is used by dermatologist most frequently. In this rule the lesion is evaluated on certain parameters: A for asymmetry, B for border, C for color, D for differential structural of the lesion. A value known as Total Dermatoscopy Score (TDS) is calculated depending up on the score given by the dermatologist on each skin lesion parameter (A, B, C, and D) and multiplied by some weight factors and finally summed over the score. If TDS is above a threshold value the lesion is a malignant one otherwise benign. In the present work different image processing technique is used for extracting features from MED-NODE image dataset where the ground truth is pre established by clinical reports given dermatologist. Multi-layer neural network is used for classification. Particle swarm optimization [5] is used for neural network hidden layer neuron number selection. In Sect. 2 of this paper, related literature in this research field is discussed. Particle swarm optimization (PSO) algorithm is discussed in Sect. 3. Details about the MED-NODE dataset used in this work and flow of this work (segmentation, feature extraction, classification and result analysis) are discussed in Sects. 4 and 5 respectively. Finally in the conclusion section importance of this work is given.

2 Related Work

Giotis et al. [6] has worked with MED-NODE dataset and achieved 81% of accuracy, 81% specificity and 80% sensitivity with total 170 MED-NODE dataset images (100 nevus images and 70 melanoma images) using total 675 extracted features per images by using Cluster-based Adaptive Metric classifier developed by them. They have preprocessed every image for removal of illumination and noise effect and then mapped each image into 50 sub-images of pixel size 15×15. A total 2250 training feature vectors (50×45) and 6250 evaluation feature vectors (50×125) are obtained from the above method. Other than MED-NODE there are few other publicly available dataset on melanoma. Dermofit dataset collected by Edinburgh Innovations center of University of Edinburg consists of large number (1300 images) of melanoma and non-melanoma images, Tan et al. [7] used this dataset for their work. Image preprocessing like image contrast enhancement, hair removal and noise filtering are performed before feature extraction. Adaptive snake approach of segmentation is done after these preprocessing steps. A total of 3914 features are extracted from each image. Finally, 1472 features are selected by Genetic algorithm before classification step by dimensionality reduction. SVM classifier gives 83% sensitivity, 89% specificity and 88% accuracy of classification in Tan et al. [7] work. McDonagh et al. [8] achieved 83.7% accuracy using 234 images whereas Ballerini et al. [9] using 960 images achieved 74% accuracy with the same dataset.

Metaheuristic algorithms like PSO are used for feature selection by several researchers. Xue et al. [10] experimented on two variations of particle swarm optimization (PSO) using the concept of non-dominated sorting and using the idea of crowding, mutation, and dominance for feature selection on 12 number of standard data set collected from UCI repository with varied number of features from 13 to 617, different class size 2–7 and varied sample sized from 32 to 4400. After comparing the result with conventional feature selection method like greedy step-wise backward selection (GSBS) and linear forward selection (LFS) they found that the PSO variants select better feature subset. Metaheuristic algorithms are used in several research works for neural network parameter optimization. Rere et al. [11] used popular MNIST and CIFAR image deep learning dataset for their study. They have used three metaheuristic algorithms, Simulated Annealing (SA), Differential Evolution (DE) and Harmony Search (HS) for the optimization of weight and bias value of all the layers in the Convolution Neural Network (CNN). They have used DeepLearn and MatConvNet Toolbox for their work. It is found from Rere et al. work, that using the three metaheuristic Simulated Annealing, Differential Evolution and Harmony Search the accuracy of the CNN improves. For example the accuracy is 5.73% greater in CNN with SA optimization, 3.91% greater in CNN with DE optimization and 7.14% greater in CNN with HS optimization than only CNN. Ojha et al. [12] has done an elaborated survey on the use of metaheuristic in neural networks design in last 20 years.

3 Particle Swarm Optimization

Particle Swarm Optimization (PSO) is a metaheuristic technique which follows the social behavior of animals like fish and bird, which is first proposed by Eberhart and Kennedy in 1995 [5]. The algorithm finds the optimized solution from better to best in consecutive iteration. Each particle in the population is a candidate solution of the optimization problem. Each individual particle has a position in the search space of all possible solution. Let particle p has a position POS_p in the search space SP in Fig. 1. The position of the pth particle is shown as $POS_p(t)$. Every particle requires memory to store the knowledge of the value (best solution, i.e., fitness function) of the personal best solution, $PBest_p(t)$, achieved by the corresponding particle and another memory keeps the global best solution, $GBest(t)$. The particle $POS_p(t)$ has

Fig. 1 a, b Malignant melanoma image   **(a)**   **(b)**

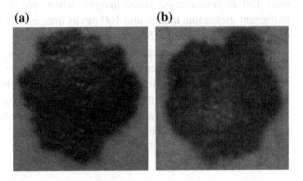

the velocity $V_p(t)$. Due to the presence of three vectors $V_p(t)$, $(PBest_p(t) - POS_p(t))$ and $(GBest(t) - POS_p(t))$ the particle moves from its initial location to its updated position.

Updating the position of the particle is done by the following equations (Eqs. 1 and 2):

$$POS_p(t + 1) = POS_p(t) + V_p(t + 1) \tag{1}$$

where

$$V_p(t + 1) = w.V_p(t) + c1.r1.\ (PBest_p(t) - POS_p(t)) + c2.r2.(GBest(t) - POS_p(t)) \tag{2}$$

where w, $c1$ and $c2$ are real valued co-efficient. In Eq. 2, w is called inertia co-efficient and $c1$, $c2$ are called acceleration co-efficient. More precisely $c1$ is the individual particle learning rate and $c2$ is the social influence. $r1$ and $r2$ are uniformly distributed random number between 0 and 1.

The step wise PSO algorithm is as follows:

Step 1: Take a group of N particle.
Step 2: Initialize position and velocity vector $POS_p(t)$ and $V_p(t)$ respectively of each particle p in time t.
Step 3: Calculate the personal best solution, $PBest_p(t)$ of each particle p and the global best $GBest(t)$ is the best among personal bests.
Step 4: Revise every particle's position vector $POS_p(t)$ and the velocity vector $V_p(t)$ by Eqs. 1 and 2.
Step 5: Execute step 3 and step 4 until the termination condition is met.
Step 6: Output the final solution.

4 MED-NODE Dataset

In this work, standard MED-NODE melanoma dataset [6] is used. This dataset has total 170 high-resolution color images, which are divided in two groups of 70 malignant melanoma image, and 100 nevus images. MED-NODE is a subset of skin disease lesion archive of 50,000 images collected by University Medical Center Groningen. These images are taken by Nikon D3 and Nikon D1x camera with a Nikkor lens from a distance of around 30 cm from the lesion area [6]. In all these JPEG, varying pixel size images lesion area can be clearly distinguishable. These images are verified and categorized by the dermatologist as gold standard [6]. In Figs. 1a, b and 2a, b two sample malignant melanoma and two nevus MED-NODE figures is shown respectively.

Fig. 2 **a, b** Nevus image   **(a)**   **(b)**

5 Present Work Flow

In this present work MATLAB (R2015b) Version 8.6.0.267246 is used for writing program of lesion image preprocessing, segmentation, feature extraction, meta-heuristic algorithm PSO and neural network based classifier. All the codes are executed in 64 bit Windows 10 operating system and Intel Core i5-6500 @ 3.20 GHz processor with 8 GB RAM.

5.1 Preprocessing and Segmentation

First step of the analysis is the segmentation of lesion area. We adopted Otsu's method [13] for this purpose which is one of the most popular cluster-based image segmentation methods. In this method, difference between the white pixel and the black pixel is taken into consideration to identify the threshold value for segmentation. Color images are converted into gray scale for segmentation. Figures 3 and 4 show the effect of segmentation of Figs. 3 and 4.

Fig. 3 **a, b** Segmented   **(a)**   **(b)**
melanoma image

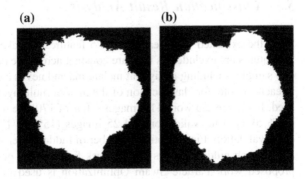

Fig. 4 a, b Segmented nevus image

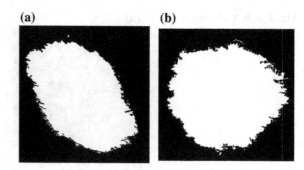

(a)   (b)

5.2 Feature Extraction

After segmentation of the lesion area by Otsu's method [13] for each of 170 MED-NODE image, total 1900 features including color, geometric shape and GLCM (Gray-Level Co-Occurrence Matrix) [14–16] and GLRLM (Gray-Level Run Length Matrix) [17] texture features are extracted. Out of these 1900 features, 36 are color features [18], 12 are geometrical shape features [7] and 1852 are texture features (88 GLRLM and 1764 GLCM). For extracting color feature, six different color models are used (RGB, HSV, YCbCr, Lab, YIQ and XYZ). In each of these color models, two features (Mean and Standard Deviation) of three different color channels are calculated. For gray scale GLRLM texture, 11 features in two different quantization levels and four different angles are extracted. In case of gray-scale GLCM, 21 features for six different inter pixel distance and two different quantization levels are extracted for only $0°$ angle. The color GLCM features are extracted for each three channels of RGB and HSV color model with two quantization levels and six inter-pixel distance with only $0°$ angle. The details of all 1900 features extracted are given in Table 1.

5.3 Classification Result Analysis

From the extracted 1900 features, 1875 features are taken for entire experiment as 25 features are excluded as they are constant across the dataset. The full dataset has 170 samples including malignant melanoma and nevus and 1875 number of features in each sample, for classification of the dataset multilayer neural network (MLP) is used. In the present work 119 images (70% of 170) are used for training, 26 images (15% of 170) for validation and 25 images (15% of 170) for testing with 3 fold cross-validation. Fixing optimal number of hidden layer and finding optimal neuron size in each hidden layer is a tricky task for complex problem like this one. In the proposed work Particle Swam Optimization is used to solve this problem. Two hidden layer are taken in the backpropagation MLP, with maximum 500 neurons in each layer.

Table 1 Details of features used in the work

Feature type and number	Extracted feature list
Geometrical shape feature [7] (total 12)	(i) Area, (ii) major axis length, (iii) minor axis length, (iv) irregularity indexA, (v) irregularity indexB, (vi) irregularity indexC, (vii) irregularity indexD, (viii) greatest diameter, (ix) shortest diameter, (x) perimeter, (xi) solidity and (xii) circularity index
Color feature [18] (total $2 \times 3 \times 6 = 36$)	Total 2 features namely, (standard deviation and mean) of 3 different color channels for each 6 color model (RGB, HSV, Lab, YCbCr, XYZ and YIQ)
GLRLM texture feature [17] (total $11 \times 2 \times 4 = 88$)	Total 11 features namely, ((i) short run emphasis, (ii) long run emphasis, (iii) high gray-level run emphasis, (iv) short run low gray-level emphasis, (v) short run high gray-level emphasis, (vi) gray-level non-uniformity, (vii) run length non-uniformity, (viii) run percentage, (ix) low gray-level run emphasis, (x) long run low gray-level emphasis and (xi) long run high gray-level emphasis) for 2 different quantization level (4 and 8) and for 4 different angles ($0°$, $45°$, $90°$, $135°$)
GLCM texture feature [14–16] (total $21 \times 6 \times 2 \times 7 = 1764$)	Total 21 features namely, ((i) autocorrelation, (ii) contrast, (iii) correlation, (iv) homogeneity, (v) maximum probability, (vi) sum of squares, (vii) sum average, (viii) Sum variance, (ix) sum entropy, (x) difference variance, (xi) difference entropy, (xii) information measure of correlation1, (xiii) information measure of correlation2, (xiv) cluster prominence, (xv) cluster shade, (xvi) dissimilarity, (xvii) energy, (xviii) entropy, (xix) inverse difference, (xx) inverse difference normalized and (xxi) inverse difference moment normalized) for 6 different inter pixel distance (1, 2, 3, 4, 5 and 6) with 2 different quantization level (4 and 8) for 7 different format (gray scale, 6 channels e.g., R, G, B of RGB and H, S, V of HSV color model) for only 1 angle ($0°$)

To find the optimal neuron size total 250,000 (500 neuron in first hidden layer × 500 neuron in the second hidden layer) combination are to be tested using exhaustive search, which is very time consuming. Using metaheuristic algorithm, PSO with only 1000 (100 iteration × 10 population) combination an accuracy of 85.9% is found with optimal 284 neurons in the first hidden layer and 132 neurons in the second hidden layer. In Fig. 5 the PSO iteration versus best cost (accuracy) curve for neural network optimized neuron selection is shown. Comparative result with other work with the same MED-NODE dataset is shown in Table 2. It is found that in all performance criteria, accuracy, sensitivity and specificity the proposed system performs better that the system proposed by Giotis et al. [6].

Fig. 5 Iteration versus accuracy (best cost) curve using PSO algorithm for neural network optimized neuron selection (best accuracy: 85.9%, optimal neuron: 1st layer: 284, 2nd layer: 132)

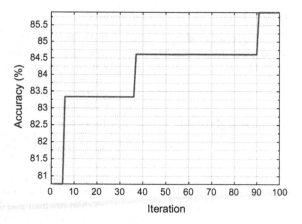

Table 2 Performance comparison of previously published best result with MED-NODE dataset

Related work	Accuracy (%)	Sensitivity (%)	Specificity (%)
Giotis et al. [6]	81	80	81
Present work	85.9	86.2	85.5

6 Conclusion

The present work achieves better result than Giotis et al. [6] with an increment of accuracy of 4.9%. In all performance criteria the present work achieves better result than Giotis et al. The present work using PSO searches optimized neural network parameter faster than exhaustive search. All extracted 1875 features are used in the classification task. Feature selection and transformation can be used to reduce feature dimensionality in future work. It is reported that the clinical accuracy is around 80% by dermatologist in diagnosis of malignant melanoma [19]. In the present work the system achieves 5.9% better accuracy than the clinical accuracy by the dermatologist for detection of malignant melanoma.

References

1. Marks, R.: Epidemiology of melanoma. Clin. Exp. Dermatol. 459–463 (2000). http://dx.doi.org/10.1046/j.1365-2230.2000.00693.x
2. Kopec, D., Kabir, M.H., Reinharth, D., Rothschild, O., Castiglione, J.A.: Human errors in medical practice: systematic classification and reduction with automated information systems. J. Med. Syst. UK **27**(4), 297–313 (2003)
3. Leo, C.D., Bevilacqua, V., Ballerini, L., Fisher, R., Aldridge, B., Rees, J.: Hierarchical classification of ten skin lesion classes (2015)

4. Rigel, D.S., Friedman, R.J.: The rationale of the ABCDs of early melanoma. J. Am. Acad. Dermatol. **29**(6), 1060–1061 (1993)

5. Kennedy, J., Eberhart, R.C.: Particle swarm optimization. In: Proceedings of the IEEE International Conference on Neural Networks IV, IEEE, Piscataway, pp. 1942–1948 (1995)

6. Giotis, I., Molders, N., Land, S., Biehl, M., Jonkman, M.F., Petkov, N.: MED-NODE: a computer-assisted melanoma diagnosis system using non-dermoscopic images. Expert Syst. Appl. **42**, 6578–6585 (2015)

7. Tan, T.Y., Zhang, L., Jiang, M.: An intelligent decision support system for skin cancer detection from dermoscopic images. In: Proceedings of the 12th International Conference on Natural Computation, Fuzzy Systems and Knowledge Discovery (ICNC-FSKD), pp. 2194–2199 (2016)

8. McDonagh, S., Fisher, R.B., Rees, J.: Using 3D information for classification of non-melanoma skin lesions. In: Proceedings of Medical Image Understanding and Analysis, pp. 164–168. BMVA Press (2008)

9. Ballerini, L., Fisher, R.B., Aldridge, B., Rees, J.: A Color and texture based hierarchical K-NN approach to the classification of non-melanoma skin lesions. In: Celebi, M., Schaefer, G. (eds.) Color Medical Image Analysis. Lecture Notes in Computational Vision and Biomechanics, vol. 6, pp. 63–86. Springer, Dordrecht (2013)

10. Xue, B., Zhang, M., Browne, W.N.: Particle swarm optimization for feature selection in classification: a multi-objective approach. IEEE Trans. Cybern. (2012)

11. Rere, L.M.R., Fanany, M.I., Arymurthy, A.M.: Metaheuristic algorithms for convolution neural network. Comput. Intell. Neurosci. Article ID 1537325 (2016). http://dx.doi.org/10.1155/2016/1537325. (Hindawi Publishing Corporation)

12. Ojha, V.K., Abraham, A., Snášel, V.: Metaheuristic design of feedforward neural networks: a review of two decades of research. Eng. Appl. Artif. Intell. **60**, 97–116 (2017)

13. Otsu, N.: A Threshold selection method from gray-level histograms. IEEE Trans. Syst. Man. Cybern. **9**(1): 62–66 (1979). https://doi.org/10.1109/tsmc.1979.4310076

14. Haralick, R.M., Shanmugam, K., Dinstein, I.: Textural features of image classification. IEEE Trans. Syst. Man Cybern. **SMC-3**(6) (1973)

15. Soh, L., Tsatsoulis, C.: Texture analysis of SAR sea ice imagery using gray level co-occurrence matrices, IEEE Trans. Geosci. Remote Sens. **37**(2) (1999)

16. Clausi, D.A.: An analysis of co-occurrence texture statistics as a function of grey level quantization. Can. J. Remote Sens. **28**(1), 45–62 (2002)

17. Tang, X.: Texture information in run-length matrices. IEEE Trans. Image Process. **7**(11), 1602–1609 (1998)

18. Pereira, S.M., Marco, A.C., Rangayyan, R.M., Azevedo-Marques, P.M.: Classification of color images of dermatological ulcers. IEEE J. Biomed. Health Inform. **17**(1) (2013)

19. Morton, C.A., Mackie, R.M.: Clinical accuracy of the diagnosis of cutaneous malignant melanoma. Br. J. Dermatol. **138**(2), 283–237 (1998). https://doi.org/10.1046/j.1365-2133.1998.02075.x

A New Search Space Reduction Technique for Genetic Algorithms

Amit Kumar Das and Dilip Kumar Pratihar

1 Introduction

Genetic algorithm (GA) is one of the most robust optimization tools used for solving a variety of problems [1]. This was introduced by Prof. Holland [2]. It mimics the basic principle of natural selection, which is nothing but the "survival of the fittest" proposed by Darwin. One cycle of a GA can be described as follows. At first, a population of solutions is created randomly and the fitness values of the solutions are evaluated. Then, a selection scheme (namely tournament selection, ranking selection, roulette-wheel selection, Boltzmann selection, etc.) is applied to select good solutions (in terms of fitness values) from the population and copy them in the mating pool. A crossover operator (such as uniform, single-point, multi-point, and others) is used to create offspring after sharing the properties of the parents. Next, the mutation operator is implemented to bring a sudden change in the solution for avoiding local minima problems. After all these operations, the fitness values of the new solutions are evaluated. At the end of the cycle, the "Elitism" principle is applied to replace the old population with a new one. In this way, the cycle, starting from the selection operation to the replacing of old solutions by new ones, goes on continuing till the stopping criterion is met.

A GA is a population-based search method and it implements the concept of natural selection based on the fitness values of the solutions. The gradient information of a problem is not used in GA and therefore, it can be applied to solve discontinuous objective functions also. It is debated that [3], though GA is a powerful tool for global optimization, it suffers from the problem of poor local

A. K. Das (✉) · D. K. Pratihar
Department of Mechanical Engineering, Indian Institute of Technology Kharagpur,
Kharagpur 721302, India
e-mail: amit.besus@gmail.com

D. K. Pratihar
e-mail: dkpra@mech.iitkgp.ac.in

© Springer Nature Singapore Pte Ltd. 2019
J. K. Mandal et al. (eds.), *Contemporary Advances in Innovative and Applicable Information Technology*, Advances in Intelligent Systems and Computing 812,
https://doi.org/10.1007/978-981-13-1540-4_12

search capability. In addition, a GA is found to have slow convergence rate, in general. To overcome these limitations of a GA, a lot of studies had been done, such as hybridizing GA with other efficient local search techniques [4], introduction of several reproduction schemes, and others. Search space reduction method is one of such techniques for improving the local search capability and convergence rate of a GA.

1.1 Search Space Reduction Techniques

Similar to other evolutionary algorithms, GA also starts with an initial population created at random in the ranges of decision variables and finally, it converges to an optimal solution after a few iterations. The diversity of the population is found to be high at the earlier iterations and it is reduced over the generations. After performing the global search, when a GA tries to converge near an optimal solution, this stage of evolution is basically a stage of refinement of the solutions, where a local search is performed. A search space reduction technique (SSRT) is intended to find the most promising area from the original variables' space and to improve the local search ability of a GA for reaching the globally optimal solution. There have been a lot of efforts made to find the said promising areas during an evolution and to confine the search process into these regions to reach the globally optimal solution at a faster rate. These efforts can be divided into two groups: decomposition of a problem [5] and search space reduction [6]. Zhao and Sannomiya [7] proposed a search space reduction method for solving large-scale flow shop problems. They developed the concept of a set of consecutively included search spaces and moreover, they proposed a modified uniform crossover operator to confine the search in the reduced regions. Barolli et al. [8] proposed a search space reduction technique for improvement of the performance of a GA-based QoS routing method in ad hoc networks. Ullah et al. [9] introduced a concept of SSRT for constrained optimization with a very narrow feasible space. In their work, they have proposed the SSRT as an initial step of the algorithm. An improved search space resizing method for model identification using a GA was proposed by Rajarathinam et al. [10]. This was applied to two processes related to third-order transfer function model. In this study, we are aiming to improve the performance of a GA with search space reduction technique. Besides several benefits of the search space reduction technique (SSRT), there have been various risks also. A reduced search region may not cover the globally optimal basin and therefore, a GA will not reach the globally optimum solution. In addition, using the search space reduction technique, the diversity of the population, as well as the exploration capability of a GA is going to be reduced and the search of the GA may not be proper. Therefore, a few queries associated with this method are to be addressed, as follows:

- What are the suitable conditions for applying SSRT?
- What will be the size of reduced search space?
- How long should we apply this SSRT?

In our proposed SSRT, we have tried to answer these questions. The rest of the paper has been organized as follows: Sect. 2 provides the detailed descriptions of the proposed SSRT. Results are stated and discussed in Sect. 3 and some conclusions are drawn in Sect. 4.

2 The Proposed Search Space Reduction Technique

For the purpose of describing the proposed search space reduction technique (SSRT), we are going to define a few terms, as follows:

Diversity Quotient (DQ): It is defined as the ratio of the absolute difference between the best fitness (f_{best}) and average fitness (f_{avg}) of the population to that between the best fitness and worst fitness (f_{worst}) values of the population.

$$DQ = \frac{|f_{best} - f_{avg}|}{|f_{best} - f_{worst}|} \tag{1}$$

This parameter is calculated at each generation and its value is found to vary between zero and one. However, if the value of f_{best} is found to be equal to f_{avg}, a small positive value (say, 1E−04) is assigned to DQ.

Maximum Diversity Quotient (DQ_{max}): It is the maximum of all the measured values of DQ and it is also calculated in each iteration.

Threshold of Diversity Quotient ($DQ_{threshold}$): It is the threshold value for the diversity quotient and is calculated using Eq. (2).

$$DQ_{threshold} = DQ_{max} \times (1 - DQ_{max}). \tag{2}$$

Probability of search space reduction (p_{ssr}): It is the probability value of search space reduction for a variable and it is determined using Eq. (3), as follows:

$$p_{ssr} = DQ_{max} \times (DQ_{max} - DQ). \tag{3}$$

Range of a variable (V_r): It is the distance between the lower and upper boundaries of a variable. For example, if a variable has lower and upper boundaries as −5 and +5, respectively, then the value of V_r will become equal to 10 for that variable.

val_{max} and val_{min}: These are the maximum and minimum values of a variable, respectively, present in the population.

At the end of each generation (except for the first generation), we are going to check, if there is any improvement in the best solution compared to that of the previous generation and accordingly, we are going to set a value for the parameter named as '*flag*'. This parameter (*flag*) will have a value of zero, if there is an improvement found in the best solution compared to that of the previous one. Otherwise, *flag* will be assigned a value equal to one. Now, the condition for applying the proposed SSRT is as follows: whenever DQ is found to be less than that of $DQ_{threshold}$ and the value of *flag* is seen to be equal to zero simultaneously, the SSRT is applied in the algorithm. The reduced search space boundaries will be calculated as follows:

$$V_{u\_r} = val_{max} + V_r \times DQ \times (1 - DQ) \times (DQ_{max} - DQ), \tag{4}$$

$$V_{l\_r} = val_{min} - V_r \times DQ \times (1 - DQ) \times (DQ_{max} - DQ), \tag{5}$$

where V_{u_r} and V_{l_r} are the upper and lower limits of a variable after reduction of the search space, respectively. If V_{u_r} is seen to be greater than the original upper limit of the variable, then V_{u_r} will be taken to be equal to the value of the original one. Similarly, the original lower limit value of a variable will be taken as the reduced lower limit value, whenever the value of V_{l_r} is found to be less than that of the original lower limit value. When the condition for SSRT is satisfied, the proposed SSRT is implemented variable-wise with a probability of p_{ssr}, that means, if a random number, created uniformly in the range of (0.0, 1.0), is found to be less than or equal to the value of p_{ssr}, the boundary limits for the variable are shortened to the values of (V_{u_r} and V_{l_r}). It is to be noted that p_{ssr} is found to be a low value at the initial stages and it increases gradually at later stages in general. This reduced search space information will be supplied to the crossover and mutation operator as the variables' boundaries and the offspring will be generated within this reduced search space.

A few important things are to be noted in this proposed SSRT. This technique is not applied from the beginning of an evolution or when the diversity of the population is found to be high. This is implemented only when there is a sufficient reduction in the diversity of the population and there is an improvement observed in the obtained best solution. It is done with the view to maintain a proper balance between the diversity and selection pressure of a GA. Moreover, the size of the reduced search space is dependent on the value of the diversity quotient (DQ), which is seen to be decreased over the generations, in general. This also helps to keep the desired balance between exploration and exploitation capabilities of a GA. Therefore, the local search capability and convergence rate of the GA are both found to be improved considerably without affecting its global search capability.

3 Results and Discussion

To measure the performance of a real-coded genetic algorithm (RCGA) with the proposed SSRT, a set of ten test functions has been taken and all these are

Table 1 Ten benchmark test functions (F01–F10)

Function name	Variable boundaries	Function name	Variable boundaries
F01: Sphere	$[-100, 100]^d$	F02: Sum of different powers	$[-100, 100]^d$
F03: Rotated hyper-ellipsoid	$[-65, 65]^d$	F04: Griewank	$[-600, 600]^d$
F05: Trid	$[-d^2, d^2]^d$	F06: Rastrigin	$[-5.12, 5.12]^d$
F07: Levy	$[-5.12, 5.12]^d$	F08: Ackley	$[-32, 32]^d$
F09: Schwefel	$[-500, 500]^d$	F10: Rosenbrock	$[-10, 10]^d$

minimization problems with 30 dimensions ($d = 30$) (refer to Table 1). The mathematical expressions for the test functions can be found in [11].

In the experiment, two real-coded genetic algorithms (RCGAs), such as an RCGA with proposed SSRT (say, RCGA_1) and a standard RCGA without SSRT (say, RCGA_2), have been used. The common evolutionary operators, which have been utilized in both the algorithms are as follows: tournament selection with a tournament size of two, simulated binary crossover [12], polynomial mutation [13] and replacement of old population with the new ones using an efficient constraint handling scheme proposed by Deb [14]. The parameters for both the RCGAs are taken the same and these are as follows: population size = 60, maximum number of generations = 500, crossover probability = 1.0, mutation probability = $1/d$, user index parameters for SBX = 2 and user index parameter for polynomial mutation = 20.

Both the algorithms have been run for 50 times for each of the test functions and the best solution available after each run has been recorded. In the experiment, the initial population for each run has been taken the same for a particular test function. From the measured results, the best, worst, mean, median and standard deviation values are calculated for each of the benchmark functions and they are given in Table 2. Moreover, an average CPU time has been observed for each of the test functions for reaching a selected objective function value (f) and these results are also provided in Table 2 (the better results are mentioned in bold). Nevertheless, it is to be noted that all the experiments have been done on an Intel i5 processor with 3.20 GHz and 16 GB RAM under Windows 10 platform and the coding of the algorithms are done using MATLAB 2017a software.

From the results, it is evident that the RCGA with the proposed search space reduction technique (i.e., RCGA_1) has given better results compared to that of the RCGA_2 for all the test functions. Moreover, the results of the time-study clearly indicate that the RCGA_1 has the faster convergence rate than that of the other one for all the cases, (as the RCGA_1 takes less CPU time to reach a particular objective function value compared to that of the RCGA_2).

As the proposed search space reduction technique has been designed in such a way that the global search capability of a GA can be maintained, while the local

Table 2 Comparison of results between RCGA_1 and RCGA_2 on test functions (F01–F10)

Function		RCGA with proposed SSRT (RCGA_1)	Standard RCGA (RCGA_2)	Avg. CPU time (in s) to reach an objective value (f)
F01	Best	**2.7564E−04**	2.9591E−03	$f = 1.6\text{E}{-}05$
	Worst	**2.2804E−03**	6.7850E−02	RCGA_1
	Mean	**9.1902E−04**	1.5450E−02	**2.3826**
	Median	**7.0261E−04**	1.1418E−02	RCGA_2
	SD	5.6543E−04	1.3758E−02	6.2401
F02	Best	**4.1832E−07**	8.3705E−06	$f = 1.6\text{E}{-}05$
	Worst	**1.5486E−03**	1.0388E−01	RCGA_1
	Mean	**1.3380E−04**	7.0785E−03	**1.0040**
	Median	**2.0488E−05**	1.0361E−03	RCGA_2
	SD	2.8650E−04	1.9230E−02	1.7233
F03	Best	**2.4586E−04**	1.8104E−02	$f = 1.0\text{E}{-}02$
	Worst	**2.0695E−02**	3.6193E−01	RCGA_1
	Mean	**5.3640E−03**	9.4360E−02	**0.7874**
	Median	**4.4068E−03**	7.9203E−02	RCGA_2
	SD	4.2029E−03	6.3707E−02	1.7164
F04	Best	**4.1429E−04**	8.1056E−03	$f = 4.0\text{E}{-}02$
	Worst	**3.8199E−02**	6.5027E−02	RCGA_1
	Mean	**1.4245E−02**	3.2367E−02	**0.6612**
	Median	**1.2541E−02**	3.0850E−02	RCGA_2
	SD	1.0014E−02	1.5091E−02	1.9588
F05	Best	**−3.5639E+03**	−2.8683E+03	$f = 9.5\text{E}{+}03$
	Worst	**2.2532E+04**	2.4498E+04	RCGA_1
	Mean	**3.4671E+03**	7.0339E+03	**0.5271**
	Median	**3.2867E+03**	4.8510E+03	RCGA_2
	SD	5.2441E+03	7.3449E+03	0.9246
F06	Best	**4.1168E−03**	1.7421E−02	$f = 9.0\text{E}{-}03$
	Worst	**2.0214E+00**	2.8311E+00	RCGA_1
	Mean	**4.1623E−01**	4.6208E−01	**0.9489**
	Median	**3.5324E−02**	8.7008E−02	RCGA_2
	SD	5.8964E−01	6.1863E−01	1.2380
F07	Best	**2.1825E−07**	2.1910E−06	$f = 2.0\text{E}{-}06$
	Worst	**6.4068E−06**	9.8977E−01	RCGA_1
	Mean	**1.4433E−06**	1.9818E−02	**1.0943**
	Median	**1.1044E−06**	1.9100E−05	RCGA_2
	SD	1.1656E−06	1.3997E−01	2.0345
F08	Best	**4.4346E−03**	1.1880E−02	$f = 1.0\text{E}{-}02$
	Worst	**1.7375E−02**	7.2418E−02	RCGA_1
	Mean	**8.7743E−03**	2.9291E−02	**0.7910**
	Median	**8.3508E−03**	2.6935E−02	RCGA_2
	SD	2.8514E−03	1.2810E−02	1.6101

(continued)

Table 2 (continued)

Function		RCGA with proposed SSRT (RCGA_1)	Standard RCGA (RCGA_2)	Avg. CPU time (in s) to reach an objective value (f)
F09	Best	**4.7376E+02**	5.9232E+02	$f = 2.0E+03$
	Worst	**1.7766E+03**	**1.7766E+03**	RCGA_1
	Mean	**1.2069E+03**	1.2308E+03	**0.1601**
	Median	**1.1844E+03**	1.2711E+03	RCGA_2
	SD	3.1317E+02	2.6686E+02	0.1840
F10	Best	**3.3546E+00**	1.5678E+01	$f = 9.0E+01$
	Worst	**1.4936E+02**	1.8732E+02	RCGA_1
	Mean	**7.0110E+01**	7.5351E+01	**0.5985**
	Median	**7.8101E+01**	7.8553E+01	RCGA_2
	SD	3.8259E+01	3.2760E+01	0.9925

search capability of the same can be improved, the RCGA with the proposed SSRT has yielded the better results compared to that of the standard RCGA (i.e., RCGA_2) in all the cases. The situation for applying the SSRT (i.e., when the algorithm already has lost a considerable amount of diversity and there has been a

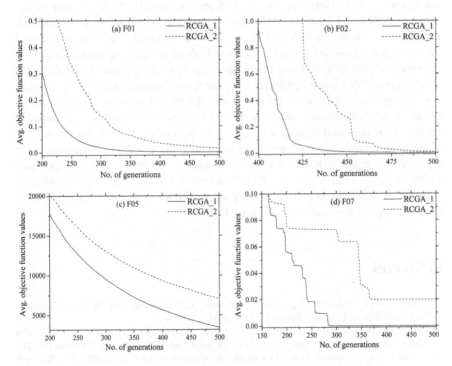

Fig. 1 Convergence graphs of the test functions: **a** F01, **b** F02, **c** F05, **d** F07

continuous improvement in the best obtained solution) has been chosen so wisely that the exploration and exploitation phenomena of the GA can be balanced properly and it will be able to perform almost equally powerful for both the global and local searches during the evolution. In addition, all the parameters, which have been used in this proposed SSRT, are calculated from the problem information and these are adaptive in nature. These inherent properties make the algorithm more robust. Nevertheless, these remarkable features of the proposed SSRT have helped the RCGA to outperform the other one.

In Fig. 1, the average of the current best objective function values of 50 runs for a particular test functions have been plotted against the number of generations for both the RCGAs. These graphs clearly show that the RCGA with the proposed SSRT has the faster convergence rate compared to that of the RCGA_2.

4 Conclusions

In this paper, a new search space reduction technique (SSRT) for a genetic algorithm has been proposed and demonstrated. The important property of this proposed SSRT is that it can maintain a proper balance between population diversity and selection pressure during an evolution. Due to this fact, local search capability and consequently, the convergence rate of a GA are improved by the application of this proposed SSRT, while the global search capability of the same is maintained. A real-coded genetic algorithm (RCGA) with the proposed SSRT has been tested on a set of ten benchmark functions and the obtained results have been compared to that of a standard RCGA. From the comparison, it is evident that the RCGA with the proposed SSRT has outperformed the other one in most of the cases. In addition, the results of the time-study reveal that the RCGA with the proposed SSRT has the faster convergence rate than that of the other one. The most remarkable feature of the proposed algorithm is that all the parameters used in this scheme are adaptive in nature and these are evaluated using the problem information. This makes the proposed SSRT more robust and suitable for use in solving a variety of problems. The similar approach will be tried for some other optimization algorithms in future. A detailed comparison of the proposed SSRT with other existing SSRTs has been kept in the scope for future study.

References

1. Golberg, D.E.: Genetic Algorithms in Search, Optimization, and Machine Learning. Addison-Wesley Longman Publishing Co., Boston (1989)
2. Holland, J.H.: Adaptation in Natural and Artificial Systems. An Introductory Analysis with Application to Biology, Control, and Artificial Intelligence. MIT Press, Cambridge (1992)
3. Pratihar, D.K.: Realizing the need for intelligent optimization tool. In: Handbook of Research on Natural Computing for Optimization Problems, IGI Global, pp. 1–9 (2016)

4. Mahmoodabadi, M.J., Safaie, A.A., Bagheri, A., Nariman-Zadeh, N.: A novel combination of particle swarm optimization and genetic algorithm for pareto optimal design of a five-degree of freedom vehicle vibration model. Appl. Soft Comput. **13**(5), 2577–2591 (2013)
5. Sannomiya, N., Iima, H., Ashizawa, K., Kobayashi, Y.: Application of genetic algorithm to a large-scale scheduling problem for a metal mold assembly process. In: Proceedings of the 38th IEEE Conference on Decision and Control, vol. 3, pp. 2288–2293 (1999)
6. Chen, S., Smith, S.F.: Improving genetic algorithms by search space reductions (with applications to flow shop scheduling). In: Proceedings of the 1st Annual Conference on Genetic and Evolutionary Computation, Morgan Kaufmann Publishers Inc., vol. 1, pp. 135–140 (1999)
7. Zhao, Y., Sannomiya, N.: An improvement of genetic algorithms by search space reductions in solving large-scale flowshop problems. IEEJ Trans. Electron. Inf. Syst. **121**(6), 1010–1015 (2001)
8. Barolli, L., Ikeda, M., De Marco, G., Durresi, A., Koyama, A., Iwashige, J.: A search space reduction algorithm for improving the performance of a GA-based qos routing method in ad-hoc networks. Int. J. Distrib. Sens. Netw. **3**(1), 41–57 (2007)
9. Ullah, B., A.S., Sarker, R., Cornforth, D.: Search space reduction technique for constrained optimization with tiny feasible space. In: Proceedings of the 10th Annual Conference on Genetic and Evolutionary Computation, Atlanta, USA, pp. 881–888 (2008)
10. Rajarathinam, K., Gomm, J.B., Yu, D., Abdelhadi, A.S.: An improved search space resizing method for model identification by standard genetic algorithm. Syst. Sci. Control Eng. **5**(1), 117–128 (2017)
11. Chakri, A., Khelif, R., Benouaret, M., Yang, X.-S.: New directional bat algorithm for continuous optimization problems. Expert Syst. Appl. **69**, 159–175 (2017)
12. Deb, K., Agrawal, R.B.: Simulated binary crossover for continuous search space. Complex Syst. **9**(3), 1–15 (1994)
13. Deb, K., Goyal, M.: A combined genetic adaptive search (GeneAS) for engineering design. Comput. Sci. Inf. **26**, 30–45 (1996)
14. Deb, K.: An efficient constraint handling method for genetic algorithms. Comput. Methods Appl. Mech. Eng. **186**(2–4), 311–338 (2000)

Part IV
Data Analytics

Reducing Computational Complexity of Skyline by the Use of Logical and Physical Bucket

Partha Ghosh, Leena Jana Ghosh, Subhajit Guha,
Narayan C. Debnath and Soumya Sen

1 Introduction

Database applications that act according to relational data model keeps the data in the form of tables and Structured Query Language (SQL) is used to fetch data from these tables. Tables in the database also can be visually represented in the form of co-ordinate system [10] where every attributes correspond to one dimension in the co-ordinate system. SQL query that are executed in a system either simple or complex in nature may be for single-criteria or multi-criteria decision making. Single-criteria decision making are easy to plan and execute. Multi-criteria decision making problems on the other hand are not possible to execute using simple SQL queries. In multi-criteria decision making often the conditions are inversely pro-

P. Ghosh (✉)
Department of Computer Application, Kingston School
of Management & Science, KEI, Kolkata, India
e-mail: pghosh44@gmail.com

L. J. Ghosh
JC Edutech, Kolkata, India
e-mail: sona.jana19@gmail.com

S. Guha
Department of Computer Science, Barrackpore Rastraguru
Surendranath College, Kolkata, India
e-mail: guha.subhajit@gmail.com

N. C. Debnath
International Society for Computers and Their Applications (ISCA),
Winona, MN, USA
e-mail: ndebnathc@gmail.com

S. Sen
A.K. Choudhury School of IT, University of Calcutta, Kolkata, India
e-mail: iamsoumyasen@gmail.com

© Springer Nature Singapore Pte Ltd. 2019
J. K. Mandal et al. (eds.), *Contemporary Advances in Innovative and Applicable Information Technology*, Advances in Intelligent Systems and Computing 812,
https://doi.org/10.1007/978-981-13-1540-4_13

portional to each other—generally in these types of cases using simple SQL queries it is not possible to execute these inverse condition relationships. In order to support inversely proportional multi-criterion decision making Skyline Operator [1, 2] has being introduced. Points those are not dominated by any other point in the system are called skyline points [1, 2]. Dominance analysis state that for two points $P1 = (p_1, p_2, p_3,..., p_d)$ and $P2 = (p_1', p_2', p_3',..., p_d')$ in d dimensional space, $P1$ dominates $P2$ if for all "i" $P1_i \leq P2_i$ where $1 \leq i \leq d$ and for at least one "j" $P1_i < P2_i$ where $1 \leq j \leq d$ [1, 2]. Hence, a point dominates another point in a d-dimensional space if it is at least equally good in all the $(d-1)$ dimensions and better in at least one dimension. In the initial conceptualization of Skyline computation [1, 2], the dominance analysis is being performed by checking the entire points and parameters. This type of computation increases time complexity. The complexity increases with increasing number of dimensions and gone up to $O(n^d)$, in which n represents number of points and d represents number of dimensions. Here we have proposed an improved concept of Skyline by incorporating spatial parameter as one of the most important dimension and performing a sorting upon the values of the spatial parameter. Now if there is a tie then the proposed methodology sorts upon the next important dimension and so on. Hence the computational time complexity reduces.

Skyline computation is widely used in travelling and tourism industries. We illustrate an example here. Generally when customers visit sea-side they prefer a hotel close to beach but cheap and the facilities like gym, restaurants etc. are available. Minimizing the distance from sea beach and reduced cost are obviously complementary conditions according to the nature of business. The challenge lies to find the alternatives which will consider multiple parameters to provide acceptable solution for the travellers.

In the proposed methodology, we perform cost analysis to map the user requirements in terms of cost. Here we categorise all the parameters of the problem domain into spatial and non-spatial parameters. The **spatial parameter** is distance and the other costs dependent on spatial parameter such as transportation costs, hotel costs etc. are **spatial dependent**. The **non spatial parameters** are gym cost, food cost etc. If someone needs to cover higher distance it increases the cost. Therefore in our proposed methodology we compute to decide which point is the best in terms of reduced cost. Hence we focus on the co-ordinate of the particular location. In order to simplify the process we transfer the reference point to its actual coordinate. Finally considering all of these we propose a methodology for multi-dimensional skyline processing that minimize the search space and hence time complexity is reduced.

The organization of this paper is as follows. In Sect. 2 gives related work, Sect. 3 describes the proposed Methodology. In Sect. 4, we have compared our proposed methodology with the other existing methods and perform a Performance Analysis. Finally in Sect. 5, we conclude and identify the future scope of this research.

2 Related Work

The first family of skyline algorithms was proposed by Borzsonyi et al. [1] and his group. A Divide-and-Conquer [1] was proposed to divide the data set into multiple sections by determining the median of the data set and in another algorithm [1] merging was shown. Both of these used sequential scanning on the dataset to compare and eliminate the dominated tuples. The complexity of these algorithms goes into $O(n (\log n)^{d-2}) + O(n \log n)$, in which n represent the number of tuples and d represent the dimension, which remains same for both the best case and worst case. A pre-sorting on tuples is being used by the SFS [3] (Sort-First-Skyline) which is actually an updated version of [1]. However Sort-First-Skyline [3] needs to scan the entire database and the time complexity remains the same in worst case. In order to avoid the scanning of whole database other approaches [5, 6] were proposed to divide the set of d-dimensional points into d sorted lists. However the complexity does not improved that much. In another research work authors have provided a comprehensive analysis and proposed a new hybrid method called Linear Elimination Sort for Skyline (LESS) [4]. The average case time complexity of LESS is $O(kn^2)$. Two algorithms, Discard_Point and BSP [10] were proposed to eliminate dominated points and find the best skyline point. The complexity of "Discard_Point" is $O(n^2)$ and of "BSP" is $O(d^2)$, where n and d are the number of points and dimension respectively. As $n \gg d$, the overall time complexity remains the same that is in the form of $O(n^2)$. Ranking neighbors by the distance parameter to a single optimal point [7–10] is the most widely studied mechanism for ranking skyline points. This mechanism first finds the optimal point in the problem domain then ranks all the skyline points according to that. The concept of computing skyline in real time [11, 12] has raised various research challenges. In many cases, dataset of real time environment may tend to be uncertain. The PSkyline [11], a work of probabilistic skyline computation has been elaborated that uses a new in-memory tree structure named as Z-tree. Another aspect of real time processing is online retrieval of skyline queries [13, 14]. Continuous evaluation of skyline over multidimensional data set in which each element is valid for a particular time range is being proposed by LookOut [14]. Another algorithm called Constrained Skyline Computing (CSC) [13] incrementally maintains all the non-redundant dominance relationships of tuples over a sliding window. However all of these algorithms are not concerned about the high computational complexity of Skyline. The worst case time complexity rises in the form of $O(n^2)$ or $O(kn^2)$. In one of the recent work [2] all the computational dimensions are transferred into one single dimension and perform a sorting upon that. This methodology is however limited as all the dimensions may not be converted into one single dimension. In this work we first identify the most important dimension that is "Spatial" dimension and converts the "Spatial Dependent" dimensions into it. Next it sums up the dimensions those are not convertible into "Spatial" dimension. Hence it restricts the problem domain into two classes. Further, the proposed methodology introduces the concept of 'Logical Bucket' and 'Physical Bucket' for avoiding the time consuming dominance analysis for computing skyline.

3 Proposed Methodology

In this research work we divide the parameters related to cost into three categories: spatial parameter, spatial dependent parameter and non spatial parameter. Distance is spatial parameter, transportation cost and hotel cost can be considered as spatial dependent parameter as transportation cost will increase if the distance is more. Generally hotel cost will decrease if we go far from the point of interest. The food cost, laundry cost and gym cost are non-spatial cost as it is different for different hotels and those are not dependent on the previously discussed dimensions. We first focus on the spatial parameter that is 'distance'. In order to simplify the problem, we consider that the reference point is at the coordinate of (0,0) from which we will chose the best object at a optimised cost. But the reference point may not be at the coordinate of (0,0), hence we map the reference point at (0,0) by transformation of coordinate and the same mapping is being applied to all the other points. Then we focus on next dimension that is 'spatial dependent' which is convertible to the 'spatial' dimension. We construct this dimension as summation of 'Hotel cost' and (distance × per unit travel cost). That is spatial dependent parameter which is computed as: Hotel cost + (distance × per unit travel cost).

We merge up these two dimensions into a single dimension to reduce complexity. The next dimension is summation of food cost, laundry cost, gym cost and other non-spatial costs. Hence we restrict the problem domain into two dimensions only, the first is the merged value of 'spatial' and 'spatial-dependent' dimension and we name it as 'sspatial' dimension and the second one is the 'non-spatial' dimension.

Thereafter we consider two buckets named logical bucket and physical bucket. The name logical suggests that after the operation, this bucket will have no existence as the bucket contains only dominated points and the name physical suggests that this bucket is permanent and will contain all non-dominating points. At first, we keep all the points into the logical bucket. Then find best points with respect to each of the two-dimensions and transfer those points into the physical bucket. Initially, with respect to the first dimension in the logical bucket compare other points one by one with the two best points in the physical bucket. The points those are not dominated with respect to two best points in the physical bucket will be added to the physical bucket and will be removed from the logical bucket. If there is a tie between the values in the 1st dimension, then we consider the 2nd dimension to break the tie. Hence after this procedure, physical bucket holds only the significant points those are the subject of interest. After getting the significant points in the physical bucket, we rank them according to the 'sspatial' dimension that is the most important 1st dimension. Now, if there is a tie in the 1st dimension, then we go for the 2nd dimension to break the tie.

Algorithm

Step 1: Start

Step 2: Map the reference point to (0,0) and change all the other point's location according to the reference point by applying transformation of coordinate geometry.

Step 3: Plot the points into d-dimensional plane and store it into a structure called 'logical_bucket'.

Step 4: Convert all the points of 'logical bucket' into 2-dimensional plane.
 Call method convert_2D(logical_bucket)

Step 5: Find the best value (minimum range) of each of the two dimensions and transfer them from Logical bucket to Physical bucket.
 Call method best_Val(Logical_bucket, Physical_bucket)

Step 6: To transfer skyline points from Logical bucket to Physical bucket
 Call method compute_Alternate_Skyline(Logical_bucket,
                                                                  Physical_bucket)

Step 7: Rank points with best prospective.
 Call method compute_rank(Physical_bucket)

Step 8: Stop

Method convert_2D(logical_bucket)
/*This method will convert all the d-dimensional points into two-dimensions.*/

Step 1: Start.

Step 2: Loop for i= 1 to logical_bucket.row_Count()
 i^{th}_point.dimension1= hotel_cost + (distance × per unit travel cost)
 i^{th}_point.dimension2= $\sum$(Non Spatial Cost)
 End for

Step 3: return updated Logical_bucket

Step 4: Stop

Method best_Val(Logical_bucket, Physical_bucket)
/* This method determine the best points for each of the 2-dimensions and transfer them from Logical_bucket to Physical_bucket. Here we use two variables position1 and position2 to hold the position of best points of each of the 2-dimensions. Min1 and Min2 are the minimum value of 1st and second dimension respectively. */

Step1: Start.

Step 2: Consider position1=1, position2=1

Step 3: Min1 = logical_bucket.point$_1$.dimension1
        Min2 = logical_bucket.point$_1$.dimension2

Step 4: Loop for i=2 To logical_bucket.row_Count()
            if(Min1 > point$_i$.dimension1) then
                    Min1= point$_i$.dimension1
                    position1=i
            end if
        End for
Step 5: Loop for i=2 To logical_bucket.row_Count()
            if(Min2 > point$_i$.dimension2) then
                    Min2= point$_i$.dimension2
                    position2=i
            end if
        End for

Step 6: Transfer points with position 'position1' and 'position2' from Logical_bucket to Physical_bucket.

Step 7: return Logical_bucket, Physical_bucket

Step 8: Stop

Method compute_Alternate_Skyline(Logical_bucket, Physical_bucket)
/* This method will transfer the non-dominating points form Logical bucket to Physical bucket. */

Step 1: Start

Step 2: Loop for i= 1 to logical_bucket.row_Count()
        if (point$_i$.dimension1 < best_val1.dimension1 AND
          point$_i$.dimension2 < best_val2.dimension2) then

            Transfer point$_i$ from Logical_bucket to Physical_bucket

      end if
      End for

Step 3: return Physical_bucket

Step 4: Stop

Method compute_rank(Physical_bucket)
/* This method performs Merge Sort upon the skyline points with respect to 1st dimension and if there is a tie in the 1st dimension, then it looks for the 2nd dimension. Finally, it will rank the skyline points.*/

Step 1: Start

Step 2: sort_w.r.t._1st_Dimension(Physical_bucket)

Step 3: if there is a tie, check 2nd dimension and change order according to
      that
Step 4: return Physical_bucket

Step 5: Stop

Table 1 Comparative study of time complexity with previous methods

Algorithm	Average case	Worst case
BNL [1]	–	$O(kn^2)$
Make_One_Dimensional, Merge sort [2]	$O(n\log_2 n)$	$O(n\log_2 n)$
SFS [3]	$O(n\log_2 n + kn)$	$O(kn^2)$
LESS [4]	$O(kn^2)$	–
Discard_Point, BSP [10]	$O(n^2)$	$O(n^2)$
CSC [13]	$O(n^2)$	$O(kn^2)$
Proposed methodology	$O(n\log_2 n)$	$O(n\log_2 n)$

4 Performance Analysis

The proposed methodology uses four algorithms namely 'convert_2D', 'best_Val', 'compute_Alternate_Skyline' and 'compute_rank'. Among those, the average case and worst case time complexity of the first three algorithms are $O(n)$ and that of the last algorithm is $O(n\log_2 n)$. Therefore the overall time complexity $= O(n) + O(n\log_2 n) \approx O(n\log_2 n)$.

In this section we have compared our proposed methodology with the previous methodologies (Table 1).

5 Conclusion and Future Works

The proposed methodology provides an effective methodology for computing and ranking Skyline Points. Main novelty of this methodology is that instead of discarding dominated points using time consuming dominance analysis, it employs the concept of logical and physical bucket with the minimum comparison. Hence the computational complexity reduces. This proposed methodology is also scalable in terms of increasing the number of dimensions.

This research work further can be extended in terms of computing skyline for multiple interesting points. In this regards, computation required to perform for multiple hotels in more than a single place. Possible extension of this research work is the reducing computational complexity for such a complex situation.

References

1. Borzsonyi, S., Kossmann, D., Stocker, K.: The skyline operator. In: International Conference on Data Engineering (ICDE) (2001)
2. Ghosh, P., Sen, S.: An alternative solution to skyline operation to reduce computational complexity. In: International Conference on Research in Computational Intelligence and Communication Networks (2016)

3. Chomicki, J., Godfrey, P., Gryz, J., Liang, D.: Skyline with presorting. In: Intelligent Information Processing and Web Mining (2005)
4. Godfrey, P., Shipley, R., Gryz, J.: Maximal vector computation in large data sets. In: Proceedings of the 31st VLDB Conference (2005)
5. Tan, K.L., Eng, P.K., Ooi, B.C.: Efficient progressive skyline computation. In: Proceedings of the International Conference on Very Large Databases (2001)
6. Eng, P.K., Ooi, B.C., Tan, K.L.: Indexing for progressive skyline computation. In: Data and Knowledge Engineering (2003)
7. Sharifzadeh, M., Shahabi, C.: The spatial skyline queries. In: Proceedings of VLDB (2006)
8. Zhang, B., Lee, K.C.K., Lee, W.C.: Location-dependent skyline query. In: Proceedings of MDM (2008)
9. Geng, M., Arefin, M.S., Morimoto, Y.: A spatial skyline for a group of user. In: International Conference on Networking and Computing (2012)
10. Ghosh, P., Sen, S.: In: Ranking skyline points by computing nearest neighbor of best skyline point. In: IEEE India International Conference (INDICON) (2015)
11. Jin, W., Tung, A. K. H., Ester, M., Han, J.: Efficient processing of subspace skyline queries on high dimensional data. In: IEEE SSDBM (2007)
12. Kim, D., Im, H., Park, S.: Computing exact skyline probabilities for uncertain databases. In: IEEE TKDE (2011)
13. Zhang, W., Lin, X., Zhang, Y., Wang, W., Yu, J.X.: Probabilistic skyline operator over sliding windows. In: IEEE International Conference Data Engineering (ICDE) (2009)
14. Jiang, B., Pei, J.: Online interval skyline queries on time series. In: IEEE International Conference on Data Engineering (ICDE) (2009)

An Integrated Blood Donation Campaign Management System

Lalmohan Dutta, Giridhar Maji, Partha Ghosh and Soumya Sen

1 Introduction

Mankind has made tremendous advancement in science and technology, unfortunately artificial blood is yet to be manufactured for the use of human beings. Fresh and safe blood is often required for several treatments most of which are life threatening and fatal in nature. Therefore, demand for blood is on the rise even more with aging population in developed countries. Authors in [1] have discussed the upcoming challenges in blood supply management with great details. Reducing wastage, proper blood inventory management, transfusing only required blood components to patients instead of giving whole blood (so that a single unit of blood can benefit multiple patients) are some of the process improvement that can help only to a limited extent. In order to support and supply the rising demand, voluntary blood is the best option. This is why blood donation camps are organized. We have to increase the blood collection in proportion to the demand otherwise we will endanger ourselves. Unfortunately, in most of the developing and under developed countries there is huge shortage of human blood supply. As per current statistics in

L. Dutta · S. Sen (✉)
A.K. Choudhury School of I.T, University of Calcutta, Kolkata, India
e-mail: iamsoumyasen@gmail.com

L. Dutta
e-mail: lalmohan.dutta@gmail.com

G. Maji
Department of Electrical Engineering, Asansol Polytechnic, Asansol, India
e-mail: giridhar.maji@gmail.com

P. Ghosh
Department of Computer Application, Kingston School of Managment
& Science, KEI, Kolkata, India
e-mail: pghosh44@gmail.com

© Springer Nature Singapore Pte Ltd. 2019
J. K. Mandal et al. (eds.), *Contemporary Advances in Innovative and Applicable Information Technology*, Advances in Intelligent Systems and Computing 812,
https://doi.org/10.1007/978-981-13-1540-4_14

India, blood requirement is around 12 million units for a population of 1.2 billion every year whereas blood collected is only 9 million units [3]. Currently when a blood donation camp is organized, it goes through different kind of difficulties which hinder the proper utilization of the campaign. Some of the difficulties are

- Camps are mainly organized only during festivals or to have political benefits or in some case when emergency occurs.
- People are sometimes not aware of the donation camp scheduled in his/her locality due to lack of proper information.
- Camp agency is not aware of the availability (like preferred time, count of interested donor) of donors. So, they are not sure about required infrastructure (like numbers of staffs, numbers of medical instruments) for the camp.
- In India there is no proper system available to record blood donation camp details, which can be used for analytical decision making in the future.

Blood donation proves to be a healthy habit that helps blood renewal. The volume of blood donation is 370–400 ml, almost 7.5% of the adult blood volume and within a period of 1–3 months, body makes it up [5]. Blood donors can be broadly categorized into three types [2] such as Voluntary, Family/Replacement and Paid Donor. The most important and continuous source of safe blood is voluntary unpaid donors. Our proposed system mainly focuses on these types of donors.

In India, generally blood donation camps are far away from the nearest blood bank or nodal agencies, so adequate medical staffs and instrumental infrastructure has to be brought from nodal to the camp location. A good estimation of possible number of donors by mining historical campaign data can help the organizer to plan for inventories well ahead. There is lot of studies on different components of a blood donation system but none has discussed about the necessity to design a coordinated and well-managed blood donation campaign management system. Hence in this paper we proposed a detailed architectural design and implementation approach for the above camp management module which could be used by the Government or NGOs to implement an integrated blood management framework.

1.1 UIDAI and AADHAAR

AADHAAR or UID (Unique Identity, denoted as UID hereafter) number is used to uniquely identify citizens in India [4]. Each Indian citizen is assigned a unique 12-digit number, known as AADHAAR number. Government has captured all biometric details such as retina scan, finger prints, photo, blood group along with other details like name, address, date of birth, sex, etc., during AADHAAR registration. Almost every aspect of a citizen can be uniquely identified by this UID number. Proposed system will have provision for integration of donors using AADHAAR.

In rest of the paper, the content of our research work is organized as follows: Sect. 2 explains a list of related research works in this field. Section 3 describes an overall system framework and database schema of the proposed work. Section 4 discusses major components of the system in details. Finally, Sect. 5 concludes the paper.

2 Related Study

A lot of studies have been done on effective blood donation systems. Researchers in [10] had interviewed 542 blood donors for attitude, motivations and belief towards donation of blood. It was found that a large number of citizen donated blood due to personal benefit from hospitals and other medical institution. They also mentioned that a good number of people are afraid of the side effects of blood donation like weight loss, sexual failure, high blood pressure, sudden death, etc. Henceforth proper awareness is required among people to get rid of those fears and showing positive attitude towards saving life. A study in [11] showed that, every year in India, there is a need of 8 million units of blood, out of which only one third is safely donated by voluntary donors. They also conducted a survey and analyzed various factor towards attitude and knowledge of blood donation among medical college students. Authors in [12] developed a web application with mobile support for managing blood donation services using SMS-based registration, alert messages, and searching a donor, etc. where people can interact without Internet connectivity. Authors in [6] developed a blood monitoring and management system to monitor blood bag storage temperature and track the blood bag location, to be used in hospitals. Authors in [7] also proposed and implemented a web based blood management system to show the registered donor real-time location using GoogleMap. This proposed system helps the patient to search nearby donors in emergency situation. In [13], the authors proposed a cloud-based application where both donors and receivers can register with their basic information like name, blood group, location, etc., and if a match occurs, both the parties will receive alert messages for quick supply of blood on demand. Researchers in [14] have done an in-depth analysis of twitter data to for donation request and response. They also proposed an automated system using social media which would help people to reduce the gap between donors and the needy patients. Authors in [8] talked about psychology of donors in reference with the retention of existing donors so that first time donor can become repetitive donor. The paper [9] developed a simulation model base on two new cutoff level policies and compares their performance for blood collection that enables comparison on the basis of shortages, wastages, and total costs. Authors in [15] talked about an intelligence system using machine learning algorithms for selecting and notifying efficient donor. The authors also proposed some optimization techniques using machine learning and decision trees for donor notification mechanism. Researchers in [16] proposed a system to predict the number of blood donors through their blood group and age using some data mining tool. Authors in [17] proposed a Geographic Information System based application to find donors in a locality in emergency situation. In paper [18], the authors developed a web based blood information system for different historical analysis.

Most of the discussed studies are isolated solutions that solve only a part of the whole. In this work we shall design an effective and integrated blood donation campaign management system, along with UID based authentication of users that can help to efficiently manage and conduct blood donation campaigns.

3 System Framework

Basic framework of the proposed system is shown in Fig. 1. It consists of Donor-related modules and blood record management modules along with common modules like login module, validation module, etc. Section 4 describes all major modules of this system in details.

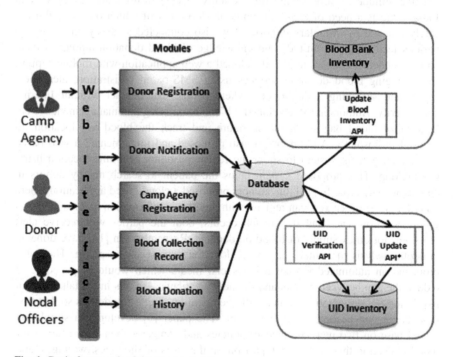

Fig. 1 Basic framework of the proposed system

3.1   Logical Data Model

Logical relational data model of our centralized database system is shown in Fig. 2. This shows the required relational tables and also relationship among them. **UserProfile** table would contain information about the different user and their roles like, donor, camp Agency, Camp organizer and System Administrator along with their login credentials. **DonorDetails** would have data about donor information like name of donor, age, sex, blood group, last donation date, AADHAAR card, etc. **CampAgencyDetails** would contain camp agency name, address, approval status, registration number, etc. **CampignDetails** will contain basic campaign related information like camp date, place, time, etc. **BloodCollectionRecord** will have all information related to each collection camp, unit of blood collected from different donors. This history would help us to derive different statistics for donors and campaign details. **RequestTracker** table will track all the requests raised by camp agency for registration and campaign to Nodal Officers. It will track till closure of the request and history will be maintained in this table.

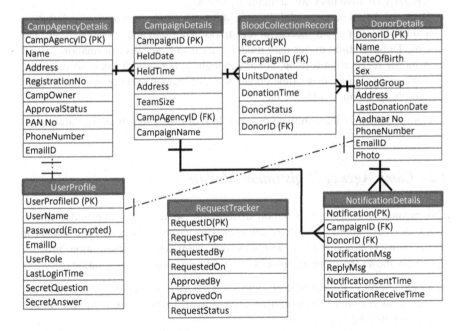

Fig. 2 Logical data model of the proposed system

4   Major Components

This section explains different major components of our proposed system. The subsections below describe the design of each important module in a generic way, which can be implemented using any web technologies.

4.1   Donor Registration Module

This module is used for registration of donor. This would be used by any user who wishes to be a blood donor. Following steps describes the donor registration process.

1. Donor enters his details like name, age, sex, blood group, AADHAAR number, locality, address, phone number and submits for registration.
2. System validates the inputs and verifies an AADHAAR UID against AADHAAR Inventory for authenticity check.
3. Once validation is successful, System stores the donor information in Database and notifies user about successful registration via email and phone message. The table **DonorDetails** in our data model is used to store this information.
4. After the registration is done, donor can login to our system using his username and password and can anytime update some of his important profile information like phone number, address, etc.

   Figure. 3a shows the complete process flow of donor registration module.

4.2   Camp Agency Registration Module

This module is used for registration of different camp agency like nodal offices, NGO, etc. This would be used by any user who wishes to be a camp organizer in our system. Following steps describes the camp agency registration process. Figure 3b shows the complete process flow of camp agency registration.

1. Camp Agency Representative enters details like name of agency, address, Registration number, Registration Date, Email, PAN number, phone number, and submits for registration.
2. After successful system validation, an email notification is sent to Nodal officers for approval of registration. Request is captured in RequestTracker table mentioned in data model. Nodal Officers log into the system and approve or reject the request with proper justification.

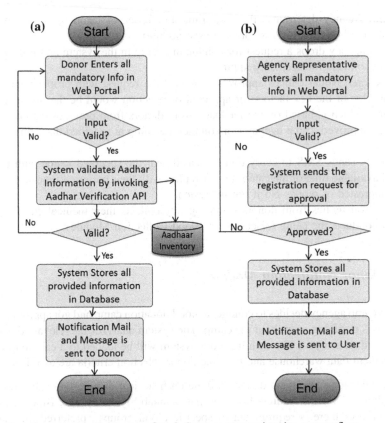

Fig. 3 **a** Donor registration process flow **b** Camp agency registration process flow

3. When registration request is approved, system stores the agency information in Database and notifies user about successful registration via email and phone message. The information is stored in CampAgencyDetails table.
4. After the registration is done, agency can login to our system using his username and password and can anytime update some of important profile information like phone number, address, etc.

4.3 Blood Campaign Management Module

When a camp agency successfully registers itself in our proposed system, the agency can anytime conduct a blood donation camp. A camp agency can be any NGO, Hospitals or other medical institutions. The Agency must take an approval from Nodal offices for each campaign before conducting. The approval process work as follows:

1. Camp Agency determines the date, time and venue of camp based on some analysis (discussed in Past data analysis module).
2. Camp agency drops a request for conducting camp in the system and it goes to Nodal officers for necessary approval.
3. Registered Nodal Offices approves or rejects the request with proper justification. One of the key factors for approval or rejection would be the organizer's rating, which is based on the feedback from donors from previous camps.
4. Once approved, camp agency can conduct the camp at specified date, time and venue.
5. Agency sends SMS to every donor located near the specified venue with date and available time of the camp. Donor replies with their preferred time slot if he is interested or replies No if not interested.
6. Based on replies from donors Camp agency collects their medical equipment and conducts the camp (discussed in Blood Collection Module).

4.4 Donor Notification Module

When a camp agency decides to conduct a blood donation camp and got approval, its first task is notifying user about the camp. The system analyzes the relevant donors with past data analysis. The data from our system will be loaded to a central blood management data warehouse and possible donor selection criteria are as follows.

1. The selection of relevant donors is done on historical data based on following attributes: Donor's Address/Locality, past availability status, age, Blood group (in case of there is requirement of specific blood groups), preferred time and Donor's response to past notifications.
2. The selected donors are notified by SMS and email in the following pattern: "*XYZ organizer is organizing a blood donation camp at Sector-5, Saltlake on 21/07/2017 at 5:00PM. Please reply your presence with YES or NO.*"
3. Interested donor will reply with YES. Donors replying with NO or no reply will be treated as NO. This information will be stored in our database for future use. Based on donors replying YES, organizer will have a prediction on how many donors may come and can take medical instruments like syringe, blood container, foods, etc., accordingly.

4.5 Blood Collection Module

On the particular day of blood camp, the organizer would setup all the prerequisite of the camp and donors will donate blood. Representative would check health status of each donor and if health status is fine to donate blood, they would allow donor to donate. Each and every donor's blood collection information would be stored in our system as follows:

1. Representative of Camp Organizer (hereby called as user) accesses the web application using browser. It can be from laptop or mobile. He must login to the application with valid credential first.
2. User will enter details like, donor id (unique Id is assigned to each donor in the system), health status, blood unit collected, time collected, etc.
3. Donor after donating blood would receive a SMS asking for feedback on the camp conducted. This would be collected in our system and organizer would be evaluated and rated using the same. This rating would be used while seeking approval for a camp.
4. After donor donates blood, two information should flow outside the system:

   a. Blood bank inventory would be updated with newly collected blood details. Proposed system would call suitable API or web service provided by centralized blood bank inventory with all details to update units of blood available.
   b. AADHAAR database can be updated through API with the date of donation and UID. This can be used to check during next donation if minimum time between consecutive donation has passed or not. This information can also be used as a privileged access card to blood from any blood bank when in need.

4.6 Blood Donation History Module

Once the blood donation is conducted, the users would be able to see the history of it for transparency. Users can be Camp Agency, Nodal Officers and Donor as well. When user logs into the system, he can see history of blood donation details, System should behave differently or show different kind of information for different users as below:

1. The Donor can see all the history of its past donations, like when where how many units of blood have been donated.
2. The Camp Agency can see all donors donated with dates, place and how many units of blood donated where conducted by the same agency. This user also can search for a particular area, donor or blood group.
3. The nodal officers can also view all details like camp agency. But nodal officers can see for all camp agencies.

   This history would help to maintain transparency in the donation system.

4.7 UID Inventory System and APIs

Our proposed framework uses UID APIs two times. First, during donor verification and second during updation of blood donation date and how many times blood donated for the donor after the donation is made. UID system already has APIs for fetching basic information of any citizen using UID number. As the system is secured, the information would be only available on the basis of One Time password (OTP). The update API is not currently available in UID inventory System. An interface needs to be provided for updating date and count to AADHAAR system. Security and Privacy are two important factors to access UID inventory through API as the inventory contains many of the private information of citizen of India. Necessary audit information can be stored in our proposed system for accountability of the person who is accessing it. If government wants to implement this integrated and centrally controlled blood campaign management system for the wellness of mankind, it must invoke the fetch and update UID API with limited access.

5 Conclusion

Detailed design and study of a blood donation camp management system with all major modules are discussed with suitable workflows. An integrated framework to manage end to end blood management process has been devised with identification of required modules, use cases, and detailed user process flows. Special focus has been given to localized blood donation campaign management where campaign activities will be planned as per the donor's preference related to time, place, date, etc. From the donor registration data camp agency can estimate the expected number of donors for a scheduled campaign and can have information on number of staff, medical instruments and other infrastructure requirement, so they can plan in advance. Furthermore, donor notification system would help to analyze and predict the responses from donor, which further can be used for analysis of different kind of donor behavior and can be used for advertisement of encouraging people for blood donation accordingly.

The future extension of this work could be data gathered through this system are stored in data warehouse for analytical processing and then future camps will be conducted based on historical data analysis. This would be more effective compared to current campaign decision triggers like political benefits or festivals or publicity as it would be on-demand requirement and dynamic in nature.

References

1. Williamson, L.M., Devine, D.V.: Challenges in the management of the blood supply. The Lancet **381**(9880), 1866–1875 (2013)
2. WHO fact sheet, http://www.who.int/mediacentre/factsheets/fs279/en/. Accessed on July 20 2017
3. India Today report, http://indiatoday.intoday.in/story/central-government-new-blood-banks-shortage/1/887649.html. Accessed on July 18 2017

4. UIDAI, http://www.uidai.gov.in. Accessed on May 1 2017
5. Blood Transfusion Council, India, http://www.mahasbtc.com. Accessed on July 20 2017
6. Kim, S.J., Yoo, S.K., Kim, H.O., Bae, H., Park, J.J., Seo, K.J., Chang, B.C.: Smart blood bag management system in a hospital environment. Lecture Notes in Computer Science, 4217, 506 (2006)
7. Ali, A., Jahan, I., Islam, A., Parvez, S.: Blood donation management system. American J. Eng. Res. **4**(6), 123–136 (2015)
8. Barbara, M., Katherine, M.W., Melissa, K.H., Deborah, J.T.: The psychology of blood donation: Current research and future directions. Transfus. Med. Rev. **22**(3), 215–233 (2008)
9. Li, B.N., Dong, M.C., Chao, S.: On decision making support in blood bank information systems. Expert Syst. Appl. **34**(2), 1522–1532 (2008)
10. Olaiya, M.A., Alakija, W., Ajala, A., Olatunji, R.O.: Knowledge, attitudes, beliefs and motivations towards blood donations among blood donors in Lagos. Nigeria Transfus. Med. **14**, 13–17 (2004)
11. Devi, H.S., Laishram, J., Shantibala, K., Elangbam, V.: Knowledge, attitude and practice (KAP) of blood safety and donation. Indian Med. Gazette **145**(1), 1–5 (2012)
12. Islam, A. S., Ahmed, N., Hasan, K., Jubayer, M.: mHealth: blood donation service in Bangladesh. In: International Conference on Informatics, Electronics & Vision (ICIEV), 17–18 May 2013
13. Pyne, B., Kundu, S., Shanmuga, S., Iyengar, N.C.S.: A smart application on cloud based blood bank. J. Comput. Math. Sci. **7**(11), 576–583 (2016)
14. Abbasi, R.A., Maqbool, O., Mushtaq, M., Aljohani, N.R., Daud, A., Alowibdi, J.S., Shahzad, B.: Saving lives using social media: analysis of the role of twitter for personal blood donation requests and dissemination. Telematics Inform. **35**(4), 892–912 (2018)
15. Chinnaswamy, A., Gopalakrishnan, G., Pandala, K.K., Venkata, S.N.: A study on automation of blood donor classification and notification techniques. Int. J. Appl. Eng. Res. **10**(7), 18503–18514 (2015)
16. Sharma, A., Gupta, P.C.: Predicting the number of blood donors through their age and blood group by using data mining tool. Int. J. Commun. Comput. Tech. **1**(6), 6–10 (2012)
17. Premasudha, B.G., Swamy, S., Adiga, B.S.: An application to find spatial distribution of blood donors from blood bank information. Int. J. Inf. Tech. Knowl. Manag. **2**(2), 401–403 (2009)
18. Khan, A.R., Qureshi, M.S.: Web-based information system for blood donation. Int. J. Dig. Cont. Tech. Appl. **3**(2), 137–142 (2009)

Comparative Study and Improvement of Various Clustering Techniques in Statistical Programming Environment

Arup Kumar Bhattacharjee, Mantrita Dey, Debalina Dutta, Sudeepa Sett, Soumen Mukherjee and Arpan Deyasi

1 Introduction

In pattern recognition literature, iris dataset is perhaps considered the best known dataset [1–4]. Nowadays work leads to huge amount of data generalization. These data are again used to many purposes. In order to ease the use to handle large data sets, data clustering is done. It is fundamental data analysis methods where groups are created by using almost similar or identical objects from a predefined set consisting of those objects; where the phrase "similarity" is utilized considering the objects in same group, and dissimilarity exists when comparison is made with other groups. Data clustering is a common technique for statistical data analysis. For this data clustering problem, R statistical programming language is used which is an open source and have immense support for statistical, machine learning, and data analysis.

A. K. Bhattacharjee (✉) · M. Dey · D. Dutta · S. Sett · S. Mukherjee
Department of Computer Application, RCC Institute of Infromation
Technology, Kolkata, India
e-mail: arupk.b@gmail.com

M. Dey
e-mail: mantrita.dey93@gmail.com

D. Dutta
e-mail: duttadebalina05@gmail.com

S. Sett
e-mail: sudeepasett@gmail.com

S. Mukherjee
e-mail: soumou601@gmail.com

A. Deyasi
Department of Electronics and Communication Engineering,
RCC Institute of Infromation Technology, Kolkata, India
e-mail: deyasi_arpan@yahoo.co.in

© Springer Nature Singapore Pte Ltd. 2019
J. K. Mandal et al. (eds.), *Contemporary Advances in Innovative and Applicable Information Technology*, Advances in Intelligent Systems and Computing 812,
https://doi.org/10.1007/978-981-13-1540-4_15

Aeberhard compared different statistical computing methods [5] where number of variables is greater than number of observations using both real and artificial data. Effective data mining technique results can be obtained through distributed multivariate regression analysis [6] on IRIS dataset. Later, data clustering is also done using unsupervised learning technique where classification accuracy is improved [7] by involving unlabeled data with unknown class. Local embedding is performed through efficient mapping procedure [8] to generate new classification techniques. k-Means algorithm is used to reduce computational load without reducing quality of findings [9]. This work is also carried out on IRIS dataset. Genetic programming is also used in classification problems in order to reduce search space size [10]. Brodley identified the requirement of normalizing the bias for feature selection [11] when dimensional factors come into effect. Kim et al. first proposed meta-evolutionary approach [12] by building small-size optimal ensembles in order to enhance the individual classifier's performance. Comparative study is made to analyze the performance between various feature-selection methods [13] when data dimensions are very high, where unvariate filter method shows better performance than multivariate technique. References [14, 15] have shown how different clustering techniques can be combined to generate better result.

In this paper, different clustering algorithms are used on IRIS dataset for the purpose of cluster analysis. 150 instances with 3 classes are used, each having 33.3% distributions. Reduction of mean is performed keeping standard deviation zero, after obtaining the xls files, and for each case, dataset is graphically represented. Here we have also shown that unclustered data coming from the result of one algorithm is treated with other algorithm which generates a hybrid environment.

2   Clustering Algorithms

In this section, we first briefly discuss the clustering algorithms used for analysis.

2.1   K-Means

K-Means, the most used nonhierarchical clustering algorithms, is used to cluster data where each cluster has a center. In K-Means the number of clusters N is considered to be fixed. Let, D is the n instance data set and let $D_1, D_2, \ldots, D_N$ be the N disjoint clusters of D. Then the error function in K-Means is defined as

$$\text{Err} = \sum_{i=1}^{N} \sum_{x \in D_i} \text{dis}(a, \mu(D_i))$$

where

$\mu(D_i)$ is the centroid of cluster D_i.

dis$(a, \mu(D_i))$ represents the distance between $\mu(D_i)$ and the data a

Mapping between a cluster D_i and the data point a is based on

$$\arg\min_{D_i \in D} \text{dist}(D_i, a)^2$$

2.2 Fuzzy K-Means

Unlike K-Means algorithm, which uses crisp partitioning, Fuzzy K-Means algorithm gives membership of each data in very cluster. For a data set $D = \{a_1, a_2, \ldots, a_n\}$, the algorithm is based on minimization of the objective function.

$$J_q(P, R) = \sum_{j=1}^{n} \sum_{i=1}^{c} \mu_{ij}^q d^2(x_j, p_i)$$

Here P is the fuzzy c-partition of the data set and R is the set of c prototypes. q is a number $(q > 1)$ which is used to *manipulate* the "fuzziness" of the clusters, p_i is the centroid of cluster i, degree of membership of object x_j belonging to cluster i is μ_{ij}, $d^2(x_j, p_i)$ represents the inner product metric and number of clusters is c.

2.3 Rough K-Means

In Rough K-Means a new feature called lower and upper approximations is used for each cluster which helps to deal with uncertainty.

There are four functions of Rough K-Means algorithms are available in R. Here we use Rough K-Means_LW on IRIS. The commonly accepted relative threshold is applied in Rough K-Means.

2.4 Fanny

In fuzzy analysis each objects i has membership for each cluster v is u_{iv}, the value of u_{iv} ranges between 1 and 0. Fanny uses fuzzy analysis for objective function minimization:

$$\sum_{v=1}^{k} \frac{\sum_{i,j=1}^{n} u_{iv}^2 u_{jv}^2 d(i,j)}{2 \sum_{j=1}^{n} u_{jv}^2}$$

where $d(i, j)$ is dissimilarities.

3 Results and Discussions

Clustering is made with 150 instances into 3 classes. Clusters are Iris Setosa (50), Iris Versicolour (50), Iris Virginica (50) with 33.3% distribution for each of 3 classes. We have applied all the clustering algorithms listed so far in Sects. 1 and 2 of this paper on IRIS dataset. Results are listed in Fig. 1.

Corresponding mean and standard deviation are listed in Table 1. For a given set of "n" no. of objects, we can set "k" clusters by using K-Means in order to achieve very high intra-cluster similarity, but simultaneously extremely low inter-cluster similarity. This algorithm is significantly sensitive when applied on the cluster centers selected at the initial phase of the problem in a random manner. Fuzzy K-Means is better than K-Means algorithm as in fuzzy analysis each objects i has membership for each cluster v (represented here as u_{iv}). In Fuzzy K-Means, for a

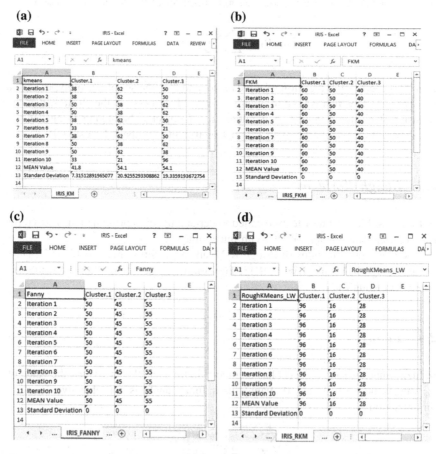

Fig. 1 Output of IRIS data applying **a** K-Means algo, **b** Fuzzy K-Means algo, **c** Fanny algo, **d** Rough K-Means LW algo

Table 1 Statistical average standard deviations of the results of IRIS dataset using all the four algorithms listed above

Algorithm	Mean			Standard deviation		
	C_1	C_2	C_3	C_1	C_2	C_3
K-Means	41.8	54.1	54.1	7.315	20.925	19.335
Fuzzy K-Means	60	50	40	0	0	0
Rough K-Means LW	96	16	28	0	0	0
Fanny	50	45	55	0	0	0

given cluster, every data points are incorporated inside it, and the degree of inclusion is determined the membership value. It is basically a technique by which data points are grouped in order to populate a given hypothetical space with n no. of dimensions into a precise number of multiple clusters. Advantage of Fuzzy K-Means is that it allows gradual membership of data points. This gives the flexibility to express the data points can belong to more than one cluster.

A vis-à-vis comparative study among different clustering techniques reveals the fact that the Fanny holds the advantage of ability of dealing with dissimilar data. The prime advantage lies in the fact that the method only deals with dissimilarities involving inter-objects, and statistical mean between the objects are not taken into account. For cluster with spherical shape, Fanny is considered as robust algorithm. Crisp clustering produced almost similar result that produced by Fanny. In that case, objects having highest membership value inside a given cluster "v" is designated by the letter "i".

Here we also applied all four algorithms ten times on the IRIS dataset. In case of Rough K-Means LW algorithm a problem occurs that few data points cannot be conclusively stored in any specific cluster. As a result sum total of data points are numerically larger than the sum total of number of data belong to that cluster, as shown in Table 2.

From the result, we can found that non-allocated data by Rough K-Means LW is 10. These non-allocated 10 data belongs to the upper bound of more than one cluster. So, they do not belong to the lower bound of any cluster. So, we cannot decide in which cluster we will store these data. After applying Fanny algorithm on those 10 non-allocated data set we get the statistical average and standard deviation which is shown in Table 3.

Now it can be easily verified from Table 3 that all the data are allocated.

Figure 2 shows number of data in each cluster for all the algorithms. From Fig. 2 we can say that after applying K-means, Fuzzy K-Means, Fanny on Iris data set we can get unique cluster for each object whereas applying RoughKMeans_LW on Iris

Table 2 Number of data points in each cluster after applying Rough K-Means LW algorithm on IRIS data

ALGO	Number of Data Points in		
	C_1	C_2	C_3
Rough K-Means LW	96	16	28

Table 3 Mean and standard deviation of data points after applying both Rough K-Means LW and FANNY Algorithm

ALGO	Mean			Standard deviation		
	C_1	C_2	C_3	C_1	C_2	C_3
Rough K-Means LW + FANNY	99	19	32	0	0	0

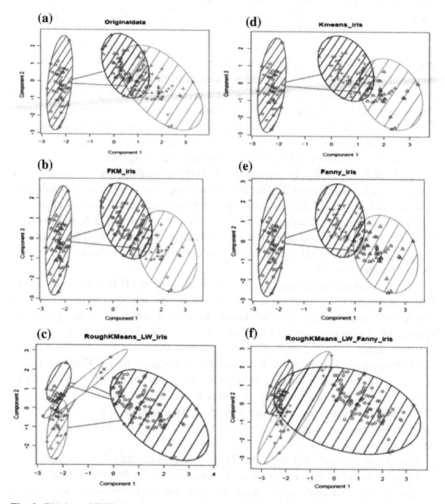

Fig. 2 Plotting of IRIS data for **a** original, **b** Fuzzy K-Means algorithm, **c** Rough K-Means LW algorithm, **d** K-Means algorithm, **e** Fanny algorithm, f Rough-K-Means LW + Fanny algorithm

we cannot find cluster for some object which are marked by symbol "×" in Fig. 2c (RoughKMeans_LW_iris). On those data points we apply Fanny and plotted in Fig. 2f (RoughKMeans_LW + Fanny).

Table 4 Mean results of PIMA, BUPA and SEED dataset using K-Means, Fuzzy K-Means, Rough K-Means LW and Fanny algorithms

Algorithm	PIMA		BUPA		SEED		
	Mean		Mean		Mean		
	C_1	C_2	C_1	C_2	C_1	C_2	C_3
K-Means	357.2	410.8	145.4	199.6	69.8	72.3	67.9
Fuzzy K-Means	518	250	53	292	77	72	61
Rough K-Means LW	518	227	312	16	69	60	47
Fanny	518	250	133	212	71	62	77

Table 5 Mean of non-conclusive data points after applying FANNY algorithm on the Rough K-Means LW clustered data set

Algorithm	PIMA		BUPA		SEED		
	Mean		Mean		Mean		
	C_1	C_2	C_1	C_2	C_1	C_2	C_3
Rough K-Means LW + FANNY	521	242	322	23	77	73	60

Similarly for few other data set, namely, PIMA Indians Diabetes Database with 768 instances into 2 classes (268 and 500 positive and negative test for diabetes), BUPA liver disorders data set with 345 instances into 2 classes (145 and 200 positive and negative cases) and SEEDS dataset with 210 instances into 3 classes (seed Kama, seed Rosa, Seed Canadian with 33.3% distribution for each of 3 classes) these algorithms are applied. The results are shown in Table 4.

Like IRIS data set, for these three data sets also Rough K-Means cannot associate all data points to any specific cluster since some point may belong to upper bound of more than one cluster. So for those non-conclusive data points FANNY algorithm is applied and the modified results are shown in Table 5.

Table 5 gives a hybrid approach to combine two or more clustering algorithm, which allocates all data points into one specific cluster. Also from Tables 1 and 4 it is found that FANNY algorithm gives better result while in two data set Fuzzy K-Means also gives result same as FANNY algorithm.

4 Conclusion

A relative study of different clustering techniques is performed in statistical computing environment like K-Means, Fuzzy K-Means, Rough K-Means, and Fanny algorithm by which we can decide which clustering technique is more efficient for a dataset. One important factor may be noted in this context that all the data are nonnegative, and there is no null value. A hybrid approach is also introduced by combining RoughKMeans_LW and Fanny algorithm. The present work can be extended in several directions such as to find the cluster of an arbitrary data of a same data set. Even these hybrid algorithms found appropriate mode of applications in case of quantitative large datasets.

References

1. Fisher, R.A.: The use of multiple measurements in taxonomic problems, Ann. Eugenics **7**(II), 179–188 (1936)
2. Duda, R.O., Hart, P.E.: Pattern Classification and Scene Analysis. Wiley, New York, p. 218 (1973)
3. Dasarathy, B.V.: Nosing around the neighborhood: a new system structure and classification rule for recognition in partially exposed environments. IEEE Trans. Pattern Anal. Mach. Intell. **PAMI-2**(1), 67–71 (1980)
4. Gates, G.W.: The reduced nearest neighbor rule. IEEE Trans. Inf. Theory **18**(3), 431–433 (1972)
5. Aeberhard, S., Coomans, D., deVel, O.: The performance of statistical pattern recognition methods. In: High Dimensional Settings. IEEE Signal Processing Workshop on Higher Order Statistics (1994)
6. Hershberger, D.E., Kargupta, H.: Distributed multivariate regression using wavelet-based collective data mining. J. Parallel Distrib. Comput. **61**(3), 372–400 (2001)
7. Demiriz, A., Bennett, K.P., Embrechts, M.J.: A genetic algorithm approach for semi-supervised clustering. Int. J. Smart Eng. Syst. Design **4**(1), 21–30 (2002)
8. Vlachos, M., Domeniconi, C., Gunopulos, D., Kollios, G., Koudas, N.: Non-linear dimensionality reduction techniques for classification and visualization. In: Proceedings of the 8th ACM SIGKDD International Conference on Knowledge Discovery and Data Mining, pp. 645–651 (2002)
9. Likas, A., Vlassis, N., Verbeek, J.J.: The global k-means clustering algorithm. Pattern Recogn. **36**(2), 451–461 (2003)
10. Eggermont, J., Kok, J.N., Kosters, W.A.: Genetic Programming for data classification: partitioning the search space. In: Proceedings of the ACM Symposium on Applied Computing, pp. 1001–1005 (2004)
11. Brodley, C.E.: Feature selection for unsupervised learning. J. Mach. Learn. Res. **5**, 845–889 (2004)
12. Kim, Y.S., Street, W.N., Menczer, F.: Optimal ensemble construction via meta-evolutionary ensembles. Expert Syst. Appl. **30**(4), 705–714 (2006)
13. Hua, J., Tembe, W.D., Koudas, N.: Performance of feature-selection methods in the classification of high-dimension data. Pattern Recogn. **42**(3), 409–424 (2009)
14. Maji, P., Pal, S.K.: Fuzzy-rough sets for information measures and selection of relevant genes from microarray data. In: IEEE Trans. Syst. Man Cybern. **40**(3), 741–752 (2010)
15. Maji, P., Paul, S.: Rough-fuzzy clustering for grouping functionally similar genes from microarray data. In: IEEE/ACM Trans. Comput. Biol. Bioinform. **10**(2), 286–299 (2013)

Context-Based Multi-document Summarization

Sheetal Sonawane, Archana Ghotkar and Sonam Hinge

1 Introduction

With rapid growth in technology, online information increased rapidly which leads information overloading. So, research in automatic text summarization has been increased to transfer information effectively. Document summarization is very useful in text classification, prediction and other information retrieval applications.

Retrieving relevant or quality sentences or features from single document or multiple documents is the process of document summarization. The quality summary should preserve meaning or theme of the document.

There are two types of methods for generation of summary [1] viz., extractive and abstractive. Significant sentences are considered in extractive summarization. Sentences are extracted based on its importance [1] with reference to the document or document collection. In opposite, abstractive summary is created based on the semantic representation of words, i.e., concepts. This summary is nearer to what human can think.

Generally summaries are created for a single document or multi-documents. In multi-document summarization [2–4], multiple co-related documents are taken for consideration to generate summary. For example the summary of news are dig out

S. Sonawane (✉) · A. Ghotkar · S. Hinge
Pune Institute of Computer Technology, Savitribai Phule Pune University, Pune, Maharashtra, India
e-mail: sssonawane@pict.edu

A. Ghotkar
e-mail: aaghotkar@pict.edu

S. Hinge
e-mail: snmhinge@gmail.com

© Springer Nature Singapore Pte Ltd. 2019
J. K. Mandal et al. (eds.), *Contemporary Advances in Innovative and Applicable Information Technology*, Advances in Intelligent Systems and Computing 812,
https://doi.org/10.1007/978-981-13-1540-4_16

in the categories like sports, entertainment, politics, etc. Summary can be generated based on user requirement. Hence, user can provide query and summary is generated based on the similarity [3] with the document. This method is known as query specific summarization.

Multi-document summarization using extractive summarization is proposed in this paper. Multi-document summarization system has increasing importance due to the need of query specific summarization. For example, in search engines, based on query the pages are suggested and summarizer returns the summary of selected pages. It definitely reduces user time.

Extractive summarization methods are largely used in the research work due to its simplicity and applicability. In this paper, Extractive summarization is proposed.

In extractive summarization, most important sentences are extracted to form final summary. The relevant document terms are usually used to represent a document. The sentence containing relevant terms are called as important sentence. Other sentences in a document closely associated with relevant sentence are extracted in extractive summary generation. Most of the current works, basic features such as term frequency, document length are used to allocate score to the document terms. In these approaches, dependency between different words is ignored. Also document term indexing is provided by giving equal weight to all document terms. In such a cases context of the document gets ignored. In large corpus, preserving document context is very important. The context of the document is nothing but the fact or circumstances in which particular event happened. In context, the presence or absence of word with another word will influence the meaning of a sentence.

Therefore, in this paper, our goal is to propose context-based-multi-document summarization. The main contributions of this research paper are

- Lexical association between terms is proposed using bi-gram model and it is used in PageRank algorithm to find out document term indexing of each input document.
- Bernoulli model of randomness is used to provide information content score in bi-gram term by considering bi-gram terms from all input documents.
- The system is evaluated using different dataset and methods. The Proposed algorithm shows better performance over existing methods.

The organization of this paper is as follows. Section 2 gives related work, Sect. 3 describes the proposed system. Results are given in Sect. 4. Further, the conclusion presented in Sect. 5.

2 Related Work

Recently, several researchers have contributed in automatic text summarization on single [5–7] and multiple text documents [2–4].

In most of the extractive summarization methods, graph-based methods are used. The graph-based algorithm is used to find most important sentence of the document. The sentences similar to important sentence are extracted in summary generation. Hence, sentence similarity calculation remains central. Hub-Authority framework [2] with features such as sentence length, cue phrase and first sentence are used for multi-document text summarization.

A vertex is considered as hub vertices labeled with features and authority vertices with the sentences. Hence good hub vertex is pointing to the good authority vertices. This relationship provides sentences with relevant information.

Extractive summary is generated by using clustering and ranking of sentences using weighted graph model [7]. Cluster-based method [8] for Chinese multi-document summarization is proposed using sentence clustering and sentence selection. Weu et al. [3] introduce the query-sensitive similarity measure into the existing graph model for sentence-sentence edge weight estimation. Researchers have been proved that Graph-based approach [9] is successful in text processing and information retrieval field.

Graphs are important to provide the structural and semantic relation [10] between nodes. A graph $G(V, E)$ is a collection of vertices and edges.

$V = \{$Sentence, Content feature, Word, Topic, Text unit$\}$ and edge gives association between vertices [3–5, 11, 12].

Weight is assigned to each edge based on different similarity methods [13] such as cosine similarity [3, 5, 6, 12], discourse similarity [7], content similarity [12], dissimilarity score [11]. Document can be represented using either directed graph [5] or un-directed graph, cyclic, or acyclic graph [5]. Summary is generated by extracting important sentences.

In graph-based methods [2–4, 7, 11, 12], Sentences are ranked using graph-based algorithm PageRank and HITS.

TexRank algorithm is proposed using graph-based approach where cosine similarity method is used to assign edge score between two sentences represented as vertices. Later ranking is applied and sentences are extracted.

Once graph is constructed, various algorithms are proposed for ranking. Hopfield network algorithm [3], the shortest path [4] are used to compute ranking.

Even Genetic Algorithm (GA) [5] is proposed which uses fitness function to express the quality of a summary such as topic relation, cohesion and readability. Contextual relations with a sentence-level attentive pooling recurrent neural network [14] are also proposed to construct context representations.

Heterogeneous graph [12] is also used which reflects the relationship between the different size of granularity nodes. Ramesh et al. introduced a graph-[11] based method which uses statistical analysis for generating summary of document.

The document can be represented using the different graphical representation based on application. The content based sentence similarity used in various approaches may lead to incorrect similarity. Based on the analysis of work done by various researchers based on different parameters such as, single document/ multi-document, vertex, and edge representation in Graph model, sentence and edge weight calculation and summary generation, there is need of enhancing summarization algorithm by analyzing context of sentence.

3 Methodolgy

Extracting significant sentences based on similarity from multiple documents is not sufficient to generate the summary.

For example, consider following sentences:

- Sentence 1: At beach Indian Prime minister Narendra Modi and Israeli Prime Minister Benjamin Netanyahu shown close friendship and shared handwritten notes.
- Sentence 2: Friends share hand written notes to show love with each other.
- Sentence 3: Netanyahu met Modi as a friend and made fun at Isreal country.

The example is represented using set as follows:

Sentence 1 = {Indian Prime minister Narendra Modi, Israeli Prime Minister Benjamin Netanyahu, friendship, beach, handwritten notes}
Sentence 2 = {love, handwritten note, friend}
Sentence 3 = {Modi, Isreal, country, fun, friend, Netanyahu}

In example, cosine similarity between sentence 1 and 2 is 12.91% while similarity between sentence 1 and 3 is 10.91%. Contextually sentence 1 and 3 are similar as it talks about same event. So, content similarity between sentences may lead to incorrect similarity of sentences.

Hence, in this research work context score is calculated based on the association among different terms.

The proposed system is shown in Fig. 1. The system contains document preprocessing, context computation and ranking and summary generation modules. The working of modules is explained in the following sections.

Document pre-processing methods such as sentence splitting, tokenization, and stop words removal are applied.

(a) Bi-gram Frequency Count

Ordered tokens are considered as input to this model. Co-occurring terms, i.e., bi-gram contains important information of context. So, finding out the association between words or terms is very important in summarization. Here, we are using

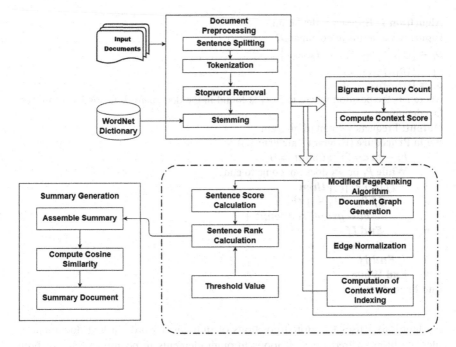

Fig. 1 Context based multi-document summarization architecture

Enhanced-Vector Space Model [15] to find out bi-gram frequency count. It is an extension to simple VSM which stores number of occurrences of tokens with its positions in the document.

$$D_j = \left\{ (t_1, P_1), (t_2, P_2), \ldots (t_N, P_N), \right\}$$

where, D_j is the vector representation of document j, t_i is the number of occurrences of the ith term in document j, N is the number of distinct terms in the document.

$$p_1 = \left\{ (p_{1i}, p_{1i+1} \ldots) \right\}$$

where i is the position where term t_1 is found in the document. k is number of places in a document.

It represents position vector of term t_1. Similarly position vector of t_2 is calculated as P_2 Based on the occurrence in the document, bi-gram is calculated. The detail algorithm is given in algorithm 1.

Algorithm 1: Bi-gram Calculation
Input: Set of terms in document$D$
$$D_j = \{(t_1, P_1), (t_2, P_2), \ldots (t_N, P_N),\}$$
$$P_1 = \{(p_{1i}, p_{1i+1} \ldots\}$$
$$P_1 = \{(p_{2i}, p_{2i+1} \ldots\}$$
where i is the position where term t_1 is found in the document. k is number of places in a document.
Output: Frequency count of bi-gram
Begin Procedure (Bi-gram_calculate(D))
  1. $bi - gramFreqCount = 0$
  2. **While** P_1 or P_2 does not come to end
  3.   **If** $|P_1| < |P_2|$ **then**
  4.   **If** $P_2 = P_1 + 1$ **then**
  5. $bi - gramFreqCount + +$
  6.   **End If**
  7. $P_1 + +, P_2 + +$
  8.   **End If**
  9. **End While**
End Procedure

Algorithm 1 uses the stored positions of individual words in text document to calculate bi-gram frequency. It loops through elements in position vector of both input elements from start and stops when any one of them reaches the end.

(b) Context Score Computation

The Bernoulli model provide a measure of the lexical association of terms from the whole corpus of documents [24].

Let D be the set of n documents.

$$D = \{D_1, D_2 \ldots D_n\}$$

T be the total set of terms in the documents.
$T = \{T_1, T_2 \ldots T_p\}$ where p is the total number of terms in all documents.
f_{ij} be the frequency of term t_j occurring in document D_i.

The document occurrence is the number of documents in which term occurs.
The document occurrence N_i is the number of documents in which term t_i occur.
The bi-gram document occurrence N_{ij} is the number of documents in which term t_i and t_j occur together.

Likelihood of term t_i appearing in document is given in Eq. 1.

$$l_i = \frac{N_i}{N} \tag{1}$$

Likelihood of occurrence of terms t_i and t_j in N_j documents is given in Eq. 2.

$$l_{ij} = \frac{N_{ij}}{N_j} \tag{2}$$

Each term in the document has some importance with respect to the document. Probability models for information retrieval based on divergence from randomness gives derivation for measuring information content in N_{ij} and N_j for term t_i and term t_j [17] is calculated using Eq. 3.

$$\begin{aligned} Inf(N_{ij}) = {} & 0.5\log_2\left(2\pi N_{ij}\left(1 - I_{ij}\right)\right) + N_{ij}\log_2\frac{I_{ij}}{I_i} \\ & + \left(N_j - N_{ij}\right)\log_2\frac{1 - I_{ij}}{1 - I_i} \end{aligned} \tag{3}$$

Information content in a bi-gram term is useful in finding the context score of bi-gram with respect to total corpus.

(c) **Sentence Score Computation**

After calculating context score, a weighted graph is created. A graph consists of nodes representing terms and edges representing context association between two terms.

Let Graph G is represented as, $G = (V, E)$, V denotes the set of terms, E is the association score.

Once graph is constructed, the relevant terms are identified using pagerank algorithm. In pagerank algorithm [18, 19] the iterations are used to normalize the algorithm, E is normalized $(\tilde{E})$ [18] by making sum of each row equal to 1.

$$\tilde{E}_{jk} = \frac{e_{jk}}{\sum_{k=1}^{|v|} e_{jk}} \tag{4}$$

Otherwise, it is 0, which is given in Eq. 4.

The context score of each word v_j in document D_i, denoted by $Context\_score(v_i)$ is calculated using algorithm 2.

Algorithm 2: Context Score Calculation
Input: $G = (V, E)$
$V = (terms \backslash bi - grams)$
$\tilde{E}_{jk}, \mu = 0.85, E = $ Context association score
Output: Context weight
Begin Procedure (Context_weight (D))
    1. $Wt[V_j] = 0$
    2.   **For each** vertex j do
    3. $Wt[V_j] \leftarrow \mu . \sum_{\forall k \neq j} Wt\,[V_k] . \tilde{E}_{kj} + \frac{1-\mu}{|v|}$
    4.   **End For**
    5.   Return Wt
End Procedure

The algorithm 2 described the procedure of assigning context weight to each document term. The sentence score is calculated on the basis of context weight of terms appeared in a sentence. Hence, Sentence weight is given in Eq. 5.

$$S_{Wt} = \sum_{j=1}^{q} Wt[t_j] \tag{5}$$

where q is number of terms in a sentence.

(d) **Summary Generation**

A sentence is a group of words. The weight of sentences is calculated by using document term weights. Sentences are ranked according to their score. Sentences with highest score are retrieved to generate a multi-document summary. To make a multi-document summary, sentences extracted from each document are combined using cosine similarity method. Computing cosine similarity has covered two major advantages that are find relevance between summary and reduce redundancy. Equation 6 provides cosine similarity between two sentences.

$$w_{\text{sim}}(s_i, s_j) = \frac{s_i \cdot s_j}{|s_i| * |s_j|} \tag{6}$$

The maximum value of $w_{\text{sim}}(s_i, s_j)$ is 1 when both sentences are identical. The final multi-document summary is shorter than original documents. Also, it gives unique information about a topic of a document.

4 Results and Discussion

Dataset and performance metric used for evaluation is explained in section (a) and result analysis is explained in section (b).

(a) **Dataset and performance metric**

Self-generated dataset (100 documents) [25], Medical dataset and the benchmark data set from the DUC 2002 [20] are used to evaluate proposed multi-document text summarization system. DUC 2002 dataset contains total 60 topics and each topic has 10 documents. The results have been evaluated using ROUGE evaluation toolkit [21], which is a n-gram based method. ROUGE stands for Recall-Oriented Understudy for Gisting Evaluation. A ROUGE N measure is given below,

$$\text{ROUGE} - N = \frac{\sum_{s \in \{\text{RefSum}\}} \sum_{n-\text{gram} \in S} \text{Count}_{\text{match}}(n\text{-gram})}{\sum_{s \in \{\text{RefSum}\}} \sum_{n-\text{gram} \in S} \text{Count}(n\text{-gram})} \quad (7)$$

(b) **Result Analysis**

To evaluate the performance of our system, we have used summary generated by the different system. There are four systems: LSA [22], TextRank [23], LexRank [16] and context-based (proposed) system. Two reference summaries are considered for analysis. For medical and 100 document summary 50% summary length is considered. For DUC 2002 dataset summary length is considered according to the reference summary length.

LSA system is based on the latent semantic analysis. It is method of extracting a set of concepts based on relationship between set of terms and document containing terms. TextRank is a graph-based algorithm and it is used to decide the importance of a vertex within the graph. The system is evaluated using single document summarization on Medical and self-generated 100 document data set. The result of performance on Medical data set is shown in Table 1.

Results on self-generated 100 document data set are shown in Table 2. The result shows that the system provides improvement in comparison with the existing system for Unigram and bi-gram matching.

Table 1 Experimental results of single document summarization on Medical data set

System	Recall	Precision	F-measure
Rouge 1			
LSA	0.405	0.631	0.466
TextRank	0.635	0.630	0.614
LexRank	0.664	0.661	0.646
Context-based	**0.874**	**0.719**	**0.776**
Rouge 2			
LSA	0.333	0.500	0.380
TextRank	0.534	0.547	0.522
LexRank	0.577	0.582	0.562
Context-based	**0.829**	**0.689**	**0.739**

The bold values indicate higher score

Table 2 Experimental results of single document summarization on self-generated (100) documents data set

System	Recall	Precision	F-measure
Rouge 1			
LSA	0.734	0.470	0.563
TextRank	0.671	**0.518**	0.567
Context-based	**0.857**	0.456	**0.584**
Rouge 2			
LSA	0.590	0.368	0.445
TextRank	0.522	0.398	0.439
Context-based	**0.790**	**0.407**	**0.527**

The bold values indicate higher score

We also have evaluated the result of multi-document summarization of system on the DUC02 data set. For evaluation, we considered first 10 categories having 10 documents in each. Four reference summaries are used for evaluation. Figures 2 and 3 shows that context-based system perform better than TextRank algorithm.

The result shows that the system provide improvement compared to TextRank algorithm. The system is also evaluated for rouge 1 to rouge 10 score given in Figs. 4 and 5 and it shows improvement over baseline algorithms.

The result analysis shows that summary obtained by the proposed system is more effective than existing systems. The performance of proposed system is increased due to following reasons:

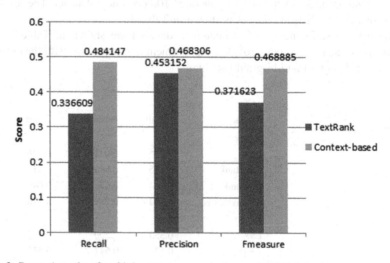

Fig. 2 Rouge 1 results of multi-document summarization on DUC02 dataset

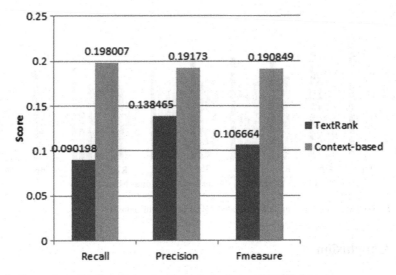

Fig. 3 Rouge 2 results of multi-document summarization on DUC02 dataset

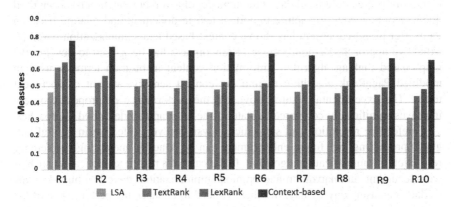

Fig. 4 Rouge 1–10 score on Medical dataset

I. The various existing system have used unigram, here we attempted to explore use of bi-gram. The bi-gram score provides the usefulness of sentence.

II. Use of lexical association between terms to provide context score. Here, document term is a vertex and context score of bi-gram is used to provide weight to the edges.

III. The vertices with highest score are extracted.

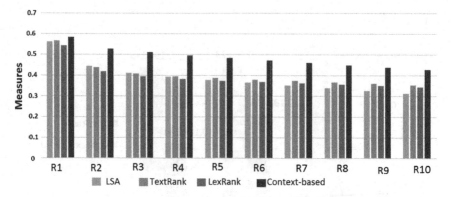

Fig. 5 Rouge 1–10 score on self-generated (100) document dataset

5 Conclusion

Automatic text summarization has great importance in today's world due to the increasingly large volume of data. Graph model of text representation has been used due to its knowledge to capture important text in document and relation between them. Our proposed system has shown context-based text summarization based on extractive approach. Bernoulli mode of randomness has been used to develop the graph-based ranking algorithm. Information content in bi-gram terms is used in modified PageRank algorithm to find out context sensitive weight of document terms. The experimental results show that the performance of proposed system improves significantly compared to state-of-the-art graph model based approaches.

Coreference resolution and semantic analysis are very effective in natural language processing. So, future enhancement can be done by integrating coreference resolution in summarization algorithm. In the case of a real-time application such as web-document summarization and opinion summarization, there is a huge volume of data increasing day by day. So, there is need to enhance the performance of the system by using distributed environment.

References

1. Das, D., Martins, A.F.T.: A survey on automatic text summarization. In: Literature Survey for the Language and Statistics II course at CMU 4, pp. 192–195 (2007)
2. Zhang, J., Sun, L., Zhou, q.: A cue-based hub-authority approach for multi-document text summarization. In: International Conference on Natural Language Processing and Knowledge Engineering, pp. 642–645 (2005)
3. Weu, F., He, Y., Li, W., Lu, Q.: A query-sensitive graph-based sentence ranking algorithm for query-oriented multi-document summarization. In: International Symposiums on Information Processing, pp. 9–13 (2008)

4. Thakkar, K.S., Dharaskar, R.V., Chandak, M.: Graph-based algorithms for text summarization. In: 3rd IEEE International Conference in Emerging Trends in Engineering and Technology (ICETET), pp. 516– 519 (2010)
5. Chatterjee, N., Mittal, A., Goyal, S.: Single document extractive texts summarization using genetic algorithms. In: Third International Conference on Emerging Applications of Information Technology (EAIT), pp. 19–23 (2012)
6. Sornil, O.. Gree-ut, K.: An Automatic text summarization approach using content-based and graph-based characteristics. In: IEEE Conference on Cybernetics and Intelligent Systems, pp. 1–6 (2006)
7. Ge, S.S., Zhang, Z., He, H.: Weighted graph model based sentence clustering and ranking for document summarization. In: 4th IEEE International Conference on in Interaction Sciences (ICIS), pp. 90–95 (2011)
8. Liu, D.-X., Hi, D.-X., Ji, D.-H., Yang, H.: A novel Chinese multi-document summarization using clustering based sentence extraction. In: Proceedings of the Fifth International Conference on Machine Learning and Cybernetics, Dalian, pp. 2592–2597 (2006)
9. Sonawane, S.S: Graph based information retrieval. IJACKD J. Res. **3**(1) (2014)
10. Sonawane, S.S., Kulkarni, P.A.: Graph based representation and analysis of text document: a survey of techniques. Int. J. Comput. Appl. **96**(19) (2014)
11. Ramesh, A., Srinivasa, K.G,, Pramod, N.: SentenceRank—a graph based approach to summarize text. In: Fifth International Conference on Applications of Digital Information and Web Technologies (ICADIWT), pp. 177–182 (2014)
12. Wei, Y.: Document summarization method based on heterogeneous graph. In: 9th IEEE International Conference on Fuzzy Systems and Knowledge Discovery, pp. 1285–1289 (2012)
13. Lin, Y.-S., Jiang, J.-Y., Lee, S.J.: A similarity measure for text classification and clustering. IEEE Trans. Knowl. Data Eng. **26**(7), 1575–1590 (2014)
14. Ren, Pengjie, Zhumin Chen, Zhaochun Ren, Furu Wei, Jun Ma, and Maarten de Rijke. Leveraging contextual sentence relations for extractive summarization using a neural attention model. In Proceedings of the 40th International ACM SIGIR Conference on Research and Development in Information Retrieval, pp. 95–104 (2017).
15. Bhakkad, A., Dharamadhikari, S.C., Kulkarni, P.: Efficient approach to find bigram frequency in text document using E-VSM. Int. J. Comput. Appl. **68** (19), 9–11 (2013)
16. Erkan, G., Ramdev, D.R.: Lexrank: graph-based lexical centrality as salience in text summarization. J. Artif. Intell. Res. 457–479 (2004)
17. Amati, G., Van Rijsbergen, C.J.: Probabilistic Models of Information Retrieval Based on Measuring the Divergence from Randomness. ACM Trans. Inf. Syst. **20**, 357–389 (2002)
18. Berberich, K., Bedathur, S., Weikum, G., Vazirgiannis, M.: Comparing Apples and oranges: normalized PageRank for evolving graphs. In: Proceedings of the 16th International Conference on World Wide Web, 1145–1146 (2007)
19. Dubey, H., Roy, B.N.: An improved page rank algorithm based on optimized normalization technique,. Int. J. Comput. Sci. Inf. Technol. **2**(5), 2183-2188 (2011)
20. Over, P., Liggett, W.: Introduction to DUC: an intrinsic evaluation of generic news text summarization systems. In: Proceedings of DUC Workshop Text Summarization (2002)
21. Lin, C.Y.: ROUGH: a package for automatic evaluation of summaries. In: Proceedings of the Workshop on Text Summarization Branches Out, (2004)
22. Steinberger, J., Jezek, K.: Using latent semantic analysis in text summarization and summary evaluation. In: Proceedings of ISIM, 93–100 (2014)
23. Mihalcea, R., Tarau, P.: Textrank: bringing order into texts. In: Proceedings of EMNLP, pp. 404–411 (2004)
24. Goyal, P., Behera, L., & McGinnity, T. M. A context-based word indexing model for document summarization. IEEE Transactions on Knowledge and Data Engineering, 25(8), 1693–1705 (2013)
25. Sonawane, S.: Extractivd Summarization dataset. Mendeley Data **1** (2018). http://dx.doi.org/10.17632/z59vy3rb2r.1

Part V
Wireless, Mobile and Cloud Computing

Probabilistic Sink Placement Strategy in Wireless Sensor Network

Krishnendu Saha, Jayanta Aich, Sumana Chakraborty
and Sayan Bose

1 Introduction

In WSN the nodes are not just detect and accumulate the information inside their working extent, yet in addition forward the information to the sink nodes [1]. The separation between typical nodes and the sink nodes are not nearer, rather the separation is higher. This prompts unequal power utilization among the sensor nodes and availability of the system might be lost [2]. Contemporary explore works show the execution, for example, information transmission time from source to sink is enhanced in different sink placement systems when contrasted and a solitary sinks placement systems. An appropriate sink arrangement system can unequivocally increment the two lifetimes of system and reductions the vitality utilization by diminishing the separation between the sensor nodes and sinks. In this paper, a probabilistic sink position approach has been actualized to put the sink in the WSNs territory keeping in mind the end goal to decrease their transmitting information time from nodes to sink, to give better vitality effectiveness and therefore to enhance organize lifetime as well as the total energy consumption.

2 Related Work

Sink Placement strategy is a challenging problem in Wireless sensor network. There are numerous sink position technique was actualized to limit the vitality utilization and to enhance sensor nodes lifetime. The exceptionally normal and commonplace sink situation procedure is Global Sink Placement Strategy (GSP). Geographical Sink Placement (GSP) [3] system puts the sinks at the focal point of gravity of area

K. Saha (✉) · J. Aich · S. Chakraborty · S. Bose
Department of CSE, Brainware University, Kolkata 700124, India
e-mail: cs.krishnendu@gmail.com

© Springer Nature Singapore Pte Ltd. 2019
J. K. Mandal et al. (eds.), *Contemporary Advances in Innovative and Applicable Information Technology*, Advances in Intelligent Systems and Computing 812,
https://doi.org/10.1007/978-981-13-1540-4_17

of a circle. Another system is Intelligent Sink Placement (ISP) [1, 4]. The ISP, the locations are updated at each sector in this methodology applicant areas are dictated by examining every single conceivable locale and relying upon the quantity of sinks, all mixes of these probable areas are counted to locate an ideal sink placement. This procedure (ISP) is observed to be an ideal one in the event of information exchange rate and vitality consumption. The other technique is Genetic Algorithm-based sink situation (GASP) [1] is additionally presented. Pant gives a decent heuristic in light of Genetic Algorithm for ideal sink placement. In sink placement [5], the issues are detailed in light of straight programming and ideal area of different sinks and information streams in WSNs are proposed. In [6], creators disclose the definition to locate the best areas of the various sink hubs and to locate the best activity stream rate are arranged.

Amplifying framework life traverse and ensure reasonableness are the significant targets of this direct programming definition. The creators arranged framework is appeared differently in relation to m-MDT (multi-sink mindful Minimum Depth Tree), and there comes about characterizes that the proposed conspire enhances organize life traverse and decency radically. The anticipated definition enables sensor hubs to speak with the at least one sink hubs amid various paths. In [7] creators plan a neighborhood look technique for sink task in WSNs that tries to lessen the greatest most pessimistic scenario delay and expand the life expectancy of a WSN. It is not suitable for a sink to utilize widespread data, which extraordinarily applies to expansive scale WSNs; they start a self-sorted out sink arrangement (SOSP) approach. The objective of this examination is to supply a superior sink task methodology with a lower correspondence overhead. By avoiding the extravagant plan of utilizing hubs area altogether, each sink keeps up its own gathering by conveying to its n-jump remove neighbors. By keeping the adjacent best residency, SOSP furnishes a prevalence of the arrangements with deference over proclamation overhead and in addition computational undertaking that are superior to prior arrangements. In [8], creators get two sink task procedures, i.e., Hopeful Location with Minimum Hop (CLMH) and Centroid of the Nodes in a Partition (CNP) and discuss their points of interest and disservices in assessment with an open policy [9]. The two procedures CLMH and CNP are coordinate to with the Geographic Sink Placement (GSP) [3] strategy which is utilized as a measuring stick. These arrangements are executed and surveyed in a diversion situation and their exhibitions are examined and examination comes about are advertised. It has been watched that the anticipated systems display enhanced exhibitions as for vitality practice and life expectancy in assessment with GSP. The creators in [2] said divergent sink situation arrangements which are Random Sink Placement (RSP) policy [10], Geographic Sink Placement (GSP) strategy, Intelligent Sink Placement (ISP), Genetic Algorithm-Based sink position (GASP) and discuss their focal points and disadvantages [9, 11]. They clarified the life expectancy of WSNs relies upon the figure of sinks with a trade stuck between the obligation cycle and the figure of sinks and additionally acquire a strategy to discover out [12].

3 Proposed Work

In this paper a new sink placement strategy has been implemented and compared with existing strategies. The proposed work is divided into problem formulation, algorithm design, and simulation work. The first requirement is to place the sink using k-means selection algorithm, then after first observation, the probability factor has been calculated according to the data transfer rate of such normal nodes and based on this factor a new iteration start and each and every iteration the probability density function get calculated and the new sink movement and placement to the best possible area order by the probability factors. At the end of the iteration the value of such data have been marked as nearest to the sink, so the data collection rates and packets to the base station have been improved. The work have been carried out using a simulation environment with a comparison based study.

4 PBSP Algorithm

Step 1: sp← Sensor Nodes position
Step 2: Tzone← Max(Zones,area)
Step 3: Select Zone heads using k-means nearest algorithm
Step 4: After the first iteration round measure the data generation rate from Zone heads
Step 5: Place the sink based on the data generation rate
Step 6: Now calculate Zone's probability using PDF for data generation rate
Step 7: Sort the Zones based on their probability factors
Step 8: Sent the packet to the base station
Step 9: Continue Step 4 and Step 5 until all dead
Step 10: End

5 Simulation Model and Results

See Table 1.

Table 1 Parameter table

Parameter	Values
Network size ($L \times L$)	100 m $\times$ 100 m
Application type	Random
Node type	Homogeneous
Transmission range (R)	5 m
Number of nodes (V)	50–150
Number of sinks (SP)	3–7
Given number of sinks (SG)	10
Packet length (l)	1024 bits
α	2
$Eelc$	50 nJ/bit/m^2
$Eamp$	10 nJ/bit/m^2

6 Simulation Results

See Figs. 1, 2 and 3.

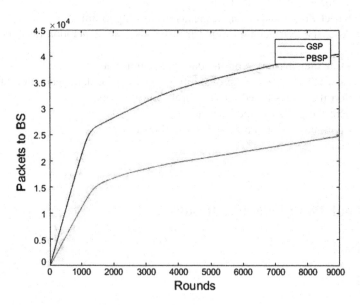

Fig. 1 Packet to base station graph

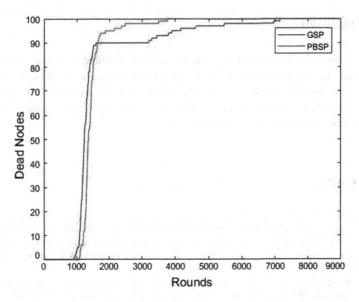

Fig. 2 Dead node calculation graph

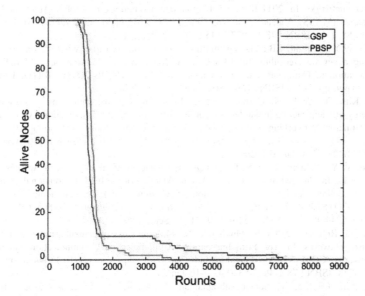

Fig. 3 Node calculation graph

7 Conclusion

Sink arrangement is a testing work henceforth requires a viable method for placement. In this printed material the proposed work were completed utilizing Matlab test system condition and contrasted and the Global Sink Placement (GSP) algorithm. Author observed that the Probability-Based Sink Placement (PBSP) works better during execution.

References

1. Chong, C.-Y.; Kumar, S.P.: Sensor networks: evolution, opportunities, and challenges. Proc. IEEE **91**(8), 1247–1256 (2003). https://doi.org/10.1109/jproc.2003.814918, ISSN-0018-9219
2. Poe, W.Y., Schmitt, J.B.: Minimizing the maximum delay in wireless sensor networks by intelligent sink placement. Technical Report 362/07. University of Kaiserslautern, Germany, July 2007
3. Vincze, Z., Vida, R., Vidács, A.: Deploying multiple sinks in multi-hop wireless sensor networks. In: IEEE International Conference on Pervasive Services (ICPS'07), Istanbul, Turkey, pp. 55–63. https://doi.org/10.1109/perser.2007.4283889, July 2007
4. Flathagen, J., Kure, Ø., Engelstad, P.E.: Constraint-based multiple sink placement for wireless sensor networks. In: 2011 IEEE 8th International Conference on Mobile Adhoc and Sensor Systems (MASS), Valencia, pp. 783–788. ISSN: 2155-6806. https://doi.org/10.1109/mass. 2011.88, Print ISBN 978-1-4577-1345-3, 17–22 Oct 2011
5. Poe, W.Y., Schmitt, J.B.: Placing multiple sinks in time-sensitive wireless sensor networks using a genetic algorithm. In: 14th GI/ITG Conference on Measurement, Modeling, and Evaluation of Computer and Communication Systems (MMB 2008), GI/ITG, Dortmund, Germany, pp. 1–15. ISBN: 978-3-8007-3090-2, Mar 2008
6. H. Kim, Y. Seok, N. Choi, Y. Choi, T. Kwon, Optimal multi-sink positioning and energy-efficient routing in wireless sensor networks. In: The 2005 International Conference on Information Networking: convergence in broadband and Mobile Networking (ICOIN'05), vol. 3391, No. 11, pp. 264–274. https://doi.org/10.1007/978-3-540-30582-8_28, ISSN-0302-9743, Jan–Feb 2005
7. Poe, W.Y., Schmitt, J.B.: Self-organized sink placement in large- scale wireless sensor networks. In: Proceedings of the 17th Annual Meeting of the IEEE International Symposium on Modeling, Analysis and Simulation of Computer and Telecommunication Systems (MASCOTS2009), London, pp. 1–3. https://doi.org/10.1109/mascot.2009.5366741. ISBN: 978-1-4244-4927-9, ISSN: 1526-7539, 21–23 Sept 2009
8. Das, D., Rehena, Z., Roy, S., Mukherjee, N.: Multiple sinks placements strategies in wireless sensor networks. In: The Fifth International Conference on Communication Systems and Networks (COMSNETS), Bangalore India, pp. 1–7. https://doi.org/10.1109/comsnets.2013. 6465578, ISBN 978-1-4673-5330-4, 7–10 Jan 2013
9. Akyildiz, I.F., Su, W., Sankarasubramaniam, Y., Cayirci, E.: Wireless sensor networks: a survey. J. Comput. Networks Int. J. Comput. Telecommun. Networking **38**(4), 393–422 (2002). https://doi.org/10.1016/S1389-1286(01)00302-4
10. Nasipuri, A., Li, K.: A Directionality based location discovery scheme for wireless sensor networks. In: Proceedings of the First ACM International Workshop on Wireless Sensor Networks and Applications (ACM WSNA 2002), Atlanta, pp. 105–111. https://doi.org/10. 1145/570738.570754, ISBN:1 58113-589-0, 28 Sept 2002

11. Rehena, Z., Roy, S., Mukherjee, N., Topology partitioning in wireless sensor networks using multiple sinks. In: The 14th IEEE International Conference on Computer and Information Technology (ICCIT 2011), Dhaka, Bangladesh, ISBN 978-1-61284- 907-2, pp. 251–256. https://doi.org/10.1109/iccitechn.2011.6164793, Dec 2011
12. Oyman, E.I., Ersoy, C.: Multiple sink network design problems in large scale wireless sensor networks. In: Proceedings of the International Conference on Communications (ICC 2004), Paris, France, vol. 6, pp. 3663–3667. https://doi.org/10.1109/icc.2004.131322, ISBN: 0-7803-8533-0, 20–24 June 2004

Continuous Monitoring of Railway Tracks with Speed Control of Rail Wirelessly

Sayan Paramanik, Krishna Sarker and Biswajit Mahanty

1 Introduction

Nowadays Railways play an important role as a medium of transport compared to planes and cars considering factors such as time, cost and capacity. But the vulnerability of train journey is sometimes major dreadful accidents caused by human blunder and ill-maintained railway tracks. Due to repetitive stresses on tracks, fatigue of track material increases day after day. To prevent accidents [1, 2] we have to ensure regular and continuous monitoring of the tracks. In our present work railway track crack detection system is slightly different from the one developed earlier [3–13]. Unlike earlier works, the Microwave [3] and Electromagnet acoustic transducer [4] are not applicable for high speed rail tracks. It is applicable for small areas and not in railways. Eddy current and ACFM [5–7] have some limitations. For eddy current testing iron fillings need to be spread over the tracks, due to large area of track it is not possible. The flow of eddy currents is always parallel to the surface and in railway tracks there is certain voltage for operating signals. So it will create problem in signalling system. Also eddy current is not applicable for large areas and complex geometries. On other side different techniques have been applied, few experiments were demonstrated using ultrasonic sensor [7–10], IR

S. Paramanik (✉) · K. Sarker
Department of Electrical Engineering, Saroj Mohan Institute
of Technology (Degree Engineering Division), Hooghly, West Bengal, India
e-mail: sayanparamanik@gmail.com

K. Sarker
e-mail: krishna80sarker@gmail.com

B. Mahanty
Department of Electronics & Communication Engineering Department,
Saroj Mohan Institute of Technology (Degree Engineering Division),
Hooghly, West Bengal, India
e-mail: mahantybiswajit@gmail.com

© Springer Nature Singapore Pte Ltd. 2019
J. K. Mandal et al. (eds.), *Contemporary Advances in Innovative and Applicable Information Technology*, Advances in Intelligent Systems and Computing 812,
https://doi.org/10.1007/978-981-13-1540-4_18

sensor [11] and LED—LDR [12]. In those techniques sensor is put with the wheel or the rail rack. When the train will be running over the track the sensor will detect if there is any crack and stop the train immediately and corresponding message will be sent to the station master. Those technologies are relevant for inspection car, not for high speed trains. Due to some practical reason this is dangerous for high speed trains. When a local train is running over the track its minimum velocity is near about 80–90 km/hr. In the same common track local, Express and Super-fast trains are also running. So it is difficult to totally stop the train at any instant of time due to high velocity, electrical and mechanical braking system and Newton's first law of motion. If we neglect these terms and stop the train at instantly then there is a chance of failure of mechanical brakes. In Indian Railways old model trains are running using DC series motor with high starting torque (New trains and engines are run by ac motor). In DC motor, there is always flashing between commutator and carbon brush, so when we stop the train at any instant of time, high flash will occur which may be the cause of burning of the motor. There are two types of coupling to connect compartments—one is Screw coupling and the other is centre buffer coupler. So when we stop instantly it may be the cause of failure of joints and there will be a chance for the train to be derailed from the track. If we put the system in inspection car, then as the car runs at very slow speed, rail traffic jam may be created. So those systems are unable to do continuous monitoring of tracks.

To overcome those problems our proposed system is designed slightly differently from the existing ones. In all experiments the track was scanned vertically [3–23] but in our present work it is scanned horizontally using ultrasonic testing. Ultrasonic testing is more efficient method to detect minor cracks and also calculate the growth rate of the crack. This system is able to do continuous monitoring of the tracks. In our work the microcontroller (Arduino) scanned the tracks before arrival of every train without any human interrogation. If the system sense any crack then it will send message to stop the train and aware the control room for public safety.

2 Existing and Proposed Techniques

2.1 Existing Technology

Due to safety reasons, now-a-days railway tracks are inspected by designated group of people or single person up to a certain region. They patrol and check the tracks on the depending on eye estimation. Sometimes they ignore or are unable to detect small or fine cracks or faulty tracks due to their limited vision. Figure 1a shows the manual track monitoring and inspection system. Another process of checking of railway tracks is done by inspection car, which is shown in Fig. 1b. Those processes are unable to perform continuous monitoring of tracks and it causes lot of rail traffic jam which is not desirable for any kind of fast communication system. Due to human limitation existing systems are time consuming, expensive and bear high chances of accidents.

(a)   (b)

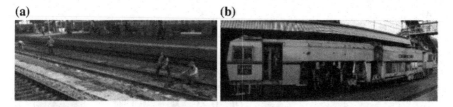

Fig. 1 **a** Manual monitoring system, **b** inspection car monitoring system

2.2 Proposed Technique

The combination of hardware and software based automatic fault detection system can identify faulty railway track using ultrasonic testing method. Also it can control the speed of rail, decrease the man power requirement with faster and automated response and its efficiency is high. Thus potentially there is rare chance of accidents plus it is cost effective. The proposed system is designed using Matlab, Arduino microcontroller and Arduino IDE 1.8.1 software.

2.3 Block Diagram of Experiment

Figure 2a, b shows the complete block diagram of the proposed system.

2.4 Simulation Diagram of the Proposed System

The hardware circuit of the proposed system is shown in Fig. 3a, b. In this work the major units are: (a) track and signalling unit and (b) Automatic speed control unit of train. Figure 4 shows the flow Chat of Track and Signalling unit.

(a)   (b)

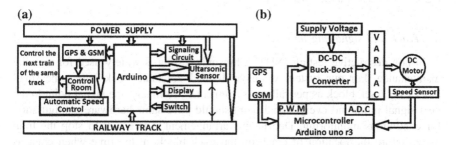

Fig. 2 Block diagram of the **a** Track and signalling, **b** speed control of train

(a)   (b)

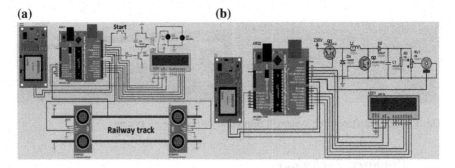

Fig. 3 Simulation circuit diagram of the **a** Track and signal, **b** automatic speed control

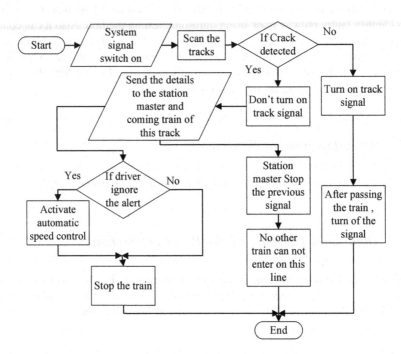

Fig. 4 Flow chart of track and signalling unit

(a) *Track and signalling unit*

In non-destructive testing (NDT) Ultrasonic testing is widely used for material maintenance. For our system we have divided the total block into some small blocks say track block as per railway systems, i.e., from any one signal point to the next signal point and every signal point should have same hardware configuration for continuous monitoring. Due to this arrangement it is easier to stop the train much before the location of crack on the track block, at the time when crack is detected. Two ultrasonic receiver and transducer are mounted on railway track as

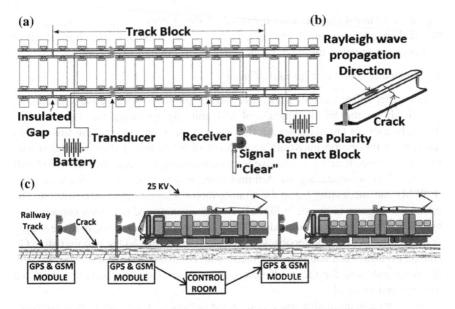

Fig. 5 Proposed configuration of the track scanning unit **a** Top view, **b** side view

shown in Fig. 5a, b which scan the track blocks horizontally with the help of microcontroller. This results a systematic monitoring of tracks depending on the scanning result, track signal will be controlled automatically. In railways with controlled signal point, the mobility and speed of rail will controlled.

At the time of train arrival into the track block, the control unit gets the start pulse from the signal operator. Then it starts monitoring the track block using ultrasonic transducer and receiver as shown in Fig. 5a, b and based on sensor output it controls the track signal. If crack is detected, then signal point does not become green, it will be Red and the details will be sent to the nearest signal operator or station master and the approaching train within a certain distance of the crack location via GSM and GPS module. If the driver by some means neglect the signal to stop the train then the automatic speed control of the train will be activated within a certain time, which will eventually stop the train. Therefore effectively the train stops before the crack and there is no possibility of any type of accident as shown in flow chart Fig. 4. The station master can also control the previous signal so that no other train can enter on this track, thereby avoiding collision. This process will be repeated continuously. The overall process will be completed within few micro-seconds, because the velocity of sound within cast iron is 5480 m/s and the microcontroller deals with micro-seconds so there is no chance of rail traffic jam.

Figure 5c demonstrates the total operation of the system. Using this configuration the track block is scanned before arrival of every train and then only it gives clearance for entering the track using signal point through microcontroller.

(b) *Automatic speed control unit of train*

For better safety another imperative unit is automatic speed control of rail. In Fig. 2b shows the Block diagram of this unit. By operating the variac, driver manually controls the motor speed. But when crack is detected and the sms will be received by the approaching train's microcontroller (Arduino), then it will first wait for driver action. If driver is inactive for a certain period of time then the speed will be controlled automatically by Arduino using pulse width modulation technique (P.W.M) and successfully stop the Train's motor without driver intervention. The schematic diagram is shown in Fig. 3b. For speed control, due to high capacity and high speed, we have incorporated a DC buck-boost converter to regulate the supply voltage using closed loop feedback control system. In Fig. 3b the two thyristors Q1 and Q2 perform buck and boost operation with controlling the firing angle. Initially Q1 is on state and the Q2 is off state, with varying the variac driver control the motor speed manually.

At the time of automatic speed control the switching operations of two thyristors Q1 and Q2 is controlled by the microcontroller depending on the rotor speed. When buck operation has to be performed then Q2 is in off state and Q1 is fired sensing the rotor speed. If the output voltage falls linearly then the microcontroller performs boost operation to maintain the speed otherwise accident might occur. At the time of buck-boost operation Q1 and Q2 both are fired depending on speed. Finally it stops the train. The overall process is shown in flowchart in Fig. 6.

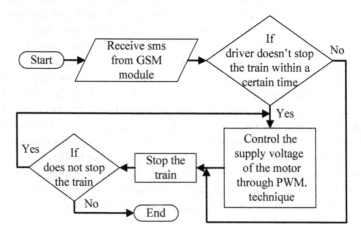

Fig. 6 Flow chart of automatic speed control

3 Modelling and Simulation for Speed Control of Train Using DC Series Motor

Due to high starting torque and load capacity DC series motor are used in rails. Figure 7a, b shows the equivalent model of the DC series motor. The armature and field resistance and inductance are in series. For speed control the inductance plays very important role. This is simulated in the Matlab simulation to determine the output response for speed control of the proposed system.

The total resistance and inductance for DC series motor are:

$$R = R_f + R_a \quad \& \quad L = L_f + L_a$$

Back Emf for DC series motor is shown in (1)

$$E_b = k_b(d\theta/dt) \tag{1}$$

From Fig. 11b the following equation govern the dynamics of an electro-mechanical system

$$\begin{aligned} V_i &= E_b + IR + L(dI/dt) \\ &= k_b(d\theta/dt) + IR + L(dI/dt) \end{aligned} \tag{2}$$

Taking Laplace transform on both sides in (2)

$$V_i(s) = k_b s\theta(s) + RI(s) + LsI(s) \tag{3}$$

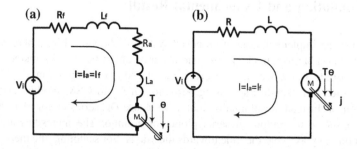

Fig. 7 DC series motor considering **a** Field and armature resistance and inductance, **b** equivalent of Fig. 7a. V_i = Input voltage, R_f = Field winding resistance, L_f = Field winding inductance, R_a = Armature winding resistance, L_a = Armature winding inductance, I = Current, T = Torque developed by motor, J = Equivalent moment of inertia of motor and load referred to the motor shaft, b = Equivalent viscous friction co-efficient of motor and load referred to the motor shaft, Θ = Angular displacement of motor shaft, k_b = Back emf constant, k_t = Motor torque constant

Maximum torque for DC series motor is in (4)

$$T = j(d^2\theta/dt^2) + b(d\theta/dt) \tag{4}$$

Taking Laplace transform on both sides in (4) and we get

$$T(s) = k_t I(s) = js^2\theta(s) + bs\theta(s)$$

$$\text{or, } I(s) = \{(js^2 + bs)\theta(s)\}/k_t \tag{5}$$

$$\text{or, } \theta(s) = k_t I(s)/(js^2 + bs) \tag{6}$$

Substituting the value of $I(s)$ in (3), we have

$$\begin{aligned} V_i(s) &= k_b s\theta(s) + \{(R + Ls)(js^2 + bs)\theta(s)/k_t\} \\ &= [k_b s + \{(R + Ls)(js^2 + bs)/k_t\}]\theta(s) \end{aligned} \tag{7}$$

The transfer function of position-voltage is

$$\theta(s)/V_i(s) = G(s)H(s) = k_t/s[(R + Ls)(js + b) + k_b k_t] \tag{8}$$

The transfer function of speed-voltage is

$$\begin{aligned} w(s)/V_i(s) &= G(s)H(s) = k_t/[(R + Ls)(js + b) + k_b k_t] \\ w(s)/V_i(s) &= G(s)H(s) = k_t/\{LJS^2 + (Lb + Rj)s + (Rb + k_b k_t)\} \end{aligned} \tag{9}$$

4 Simulation and Experimental Result

The hardware implementation of proposed system is shown in Fig. 8 here we used the rolled iron bar as a track (7 cm), and generate an artificial crack inside the bar. The ultrasonic transducer and receiver are placed according as in Fig. 5a. We implemented only single track using Arduino, GPS and GSM module and ultrasonic sensor. Instead of rail traction motor we us use DC series motor of 1500 rpm. Figure 9 shows the output of serial plotter of arduino. The intensities are being plotted on the y-axis and the time towards the x-axis. For scanning, the theoretically value $(7 \times 10^{-2}/5480)$ is 0.12 µs. which match with simulation result. At time '0' the ultrasonic transmitter sends the pulse and at time '0.12' the pulse receives at receiver side with lower amplitude.

(a)   (b)

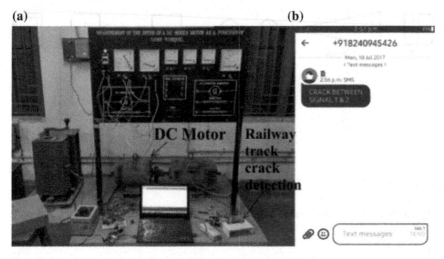

Fig. 8 Hardware setup and result **a** Different part of the hardware implementation, **b** result of GPS

Fig. 9 Amplitude versus time output for scan the track at faulty condition

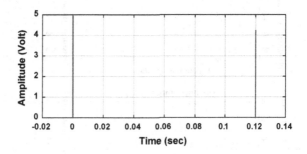

The simulation parameters are: $j = 0.0465$ kg m^2/s^2, $b = 0.00005$ (nm/rad-s^{-1}), $k_b = 0.01$, $K_t = 0.01$, $R = 205.5$ Ω, $L = 0.032$ H

From (8), we get

$$w(s)/V_i(s) = 0.01/(0.001488s^2 + 9.5565s + 0.01038) \tag{10}$$

Figure 10 shows the simulation parameter and state space diagram of the speed—voltage transfer function (10).

From (10) the calculated value of ζ is 477.8, i.e. overdamped systems. The values of ζ also changes with change of the constant parameters of the system. Figure 11 shows bode plot stability for different values of ζ. For underdamped system we consider $\zeta = 0.5$, i.e. $\zeta < 1$. The practical setup is made to see the output waveform in Matlab by function calling (with the help of 'fopen') from the Arduino uno r3. Due to high starting torque DC series motor cannot operate in no load

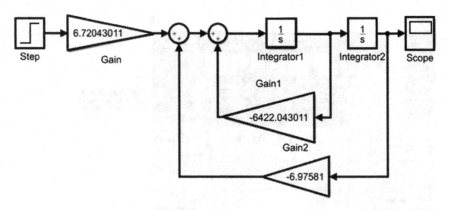

Fig. 10 State space of evaluated transfer function

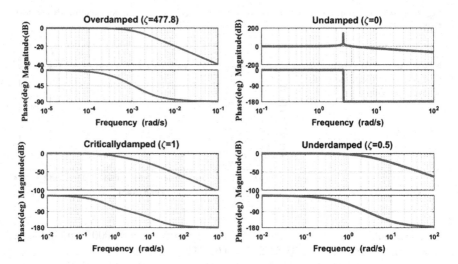

Fig. 11 DC series motor transfer function output response for different values of ζ

condition. Figure 12a, b shows the simulation result of DC series motor for load torque 100 nm and the variation of current and speed with respect to time. In Fig. 12c, d we observe the practical phenomena at faulty condition to stop the motor. Figure 12c shows the speed versus time curve. This also indicates how the speed is controlled with respect to time. Figure 12d shows the amplitude versus time response of DC series motor for load torque 100 nm and the variation of voltage using power electronics devices for speed control.

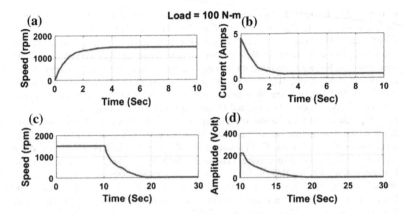

Fig. 12 Variation of **a** Speed, **b** current at normal operation with respect to time and variation of, **c** speed, **d** voltage with respect to time at faulty condition

5 Conclusions

The simulation results strongly agree with the theoretical results and this new innovative system will increase the reliability of the safety mechanism in railway transport. Also this system will be implemented in large scale to provide better safety for high speed railway tracks. There are many advantages with the proposed system compared to existing systems. The advantages are less analysis time or processing time, higher accuracy, better performance, minimum cost and uninterrupted work (at day and night) with higher efficiency. With the help of this innovative system, it is expected that, major train mishaps can be prevented and precious human life will be saved.

References

1. List of Rails Accidents in India: https://en.wikipedia.org/wiki/list_of_Indian_rail_accidents
2. A Report on Derailing 14 Coaches for Fracture Tracks: http://www.hindustantimes.com/india-news/kanpur-cracked-tracks-may-have-caused-train-to-derail-railway-orders-probe/story-BGv77L0pquIeihgjQaIvxN.html
3. Manacorda, G, Simi, A.: Non-destructive inspection and characterization of track bed with microwaves, In: 14th IEEE International Conference on Ground Penetrating Radar (GPR) Shanghai, China (2012)
4. Boughedda, H., Hacib, T., Chelabi, M., Acikgoz, H., Le Bihan, Y.: Electromagnetic acoustic transducer for cracks detection in conductive material, In: 4th IEEE International Conference on Electrical Engineering, Boumerdes, Algeria (2015)
5. Peng, J., Tian, G., Wang, L., Gao, X., Zhang, Yu., Wang, Z.: Rolling contact fatigue detection using eddy current pulsed thermography. In: Non-destructive Evaluation/Testing, IEEE, Chengdu, China (2014)

6. Kim, J., Udpa, L., Kim, J.: Development and application of Arduino based scanning system for non-destructive testing. Int. J. Appl. Eng. Res. **11**(14), 8217–8220 (2016)
7. Rowshandel, H., Icholson, G.L., Davis, C.L., Roberts, C.: A robotic system for non-destructive evaluation of RCF cracks in rails using an ACFM sensor, In: 5th IET conference on Railway condition monitoring and non-destructive testing, Derby, UK (2011)
8. Lv, W., Wu, X., Zhang, D., Ye, Y.: Quantitative estimation to circular holes in steel for non-destructive testing using ultrasonic phased array. In: IEEE International Conference on Piezoelectricity, Acoustic Waves, and Device Applications, China (2015)
9. Lad, P., Pawar, M.: Evolution of railway track crack detection system. In: 2nd IEEE International Symposium on Robotics and Manufacturing Automation, Malaysia (2016)
10. Tsunashima, H., Naganuma, Y., Matsumoto, A., Mizuma, T., Mori, H.: Condition Monitoring of Railway Track Using In-Service Vehicle. J. Mech. Syst. Transp Logistics. **3**(1) (2010)
11. Umamaheswari, Er. K., Rajesh, Er. P.: An Arduino based method for detecting cracks and obstacles in railway tracks. In: 2nd International Conference on Research Trends in Engineering, Applied Science and Management at National Institute of Technical Teachers Training & Research, Chandigarh, India (2017)
12. Muralidharan, V., Dinesh, V., Manikandan, P.: An enhanced crack detection system for railway track. Int. J. Eng. Trends Technol. **21**(6) (2015); Coimbatore
13. Kalkundri, P., Deshpande, S.: Low cost rail crack inspection system. Int. J. Mod. Trends Eng. Res. **01** (2014)
14. Orhan, S.: Analysis of free and forced vibration of a cracked cantilever beam. NDT and E Int. **40**, 443–450 (2007)
15. Nikitin, P.V., Rao K.V.S., Lazar S.: An overview of near field UHF RFID. In: IEEE International Conference on RFID, Grapevine, TX, USA (2007), 167–74
16. Tzanakakis, K.: The Railway Track and Its Long Term Behaviour, vol. 02, Springer, Berlin (2013)
17. Molodova, M., Li, Z., Núñez, A., Dollevoet, R.: Monitoring the railway infrastructure: detection of surface defects using wavelets. In: 16th International IEEE Annual Conference on Intelligent Transportation Systems, Netherlands (2013), pp. 1316–1321
18. Zhao, X., Dynamic wheel/rail rolling contact at singular defects with application to squats (Ph.D.) Delft University of Technology (2012)
19. Carrascal, I.A., Casado, J.A., Polanco, J.A., Gutierrez-Solana, F.: Dynamic behaviour of railway fastening setting pads. Eng. Fail. Anal. **14**(2), 364–373 (2007)
20. Maes, J., Sol, H., Guillaume, P.: Measurements of the dynamic railpad properties. J. Sound Vib. **293**, 557–565 (2006)
21. Kassa, E., Nielsen, J.: Dynamic interaction between train and railway turnout: full-scale field test and validation of simulation models, **46** 521–534 (2008)
22. Hernandez, G., Alberto, E.: Wheel and Rail Contact Simulation Using a Twin Disc Tester. Department of Mechanical Engineering University of Sheffield, UK (2009)
23. Ringsberg, J.W.: Life prediction on rolling contact fatigue. Int. J. Fatigue **23**, 575–586 (2001)

Secure Anonymous Session Key Agreement Between Trusted Users in Global Mobility Network

Prasanta Kumar Roy, Sathi Ball and Krittibas Parai

1 Introduction

The continuous advancement of mobile devices enables any mobile user (MU) to access the various services of *GLOMONET* [1] in a foreign agent (FA) environment via its home agent (HA). Moreover, such environments are susceptible to several security threats such as masquerading, eavesdropping and unauthorized alteration of messages [2]. Hence, a well organized communication protocol is highly required to stand against these security vulnerabilities. Recently, several authentication schemes have been proposed by many researchers to establish a session key (SK) between trusted MU and FA. Unfortunately, some of them still suffer from several disadvantages. Probably, the first SC based anonymous protocol for wireless communication using public key cryptosystem proposed by Zhu et al. [3] in the year of 2004. Lee et al. [4] claimed that Zhu et al.'s scheme cannot accomplish perfect backward secrecy, proper mutual authentication and resistance against forgery attack. They also proposed an improved technique to fulfill these shortcomings in 2006. Wu et al. [5], in 2008, found that that Lee et al.'s scheme does not achieve user anonymity and backward secrecy. They also proposed an enhanced construction to overcome these limitations. Mun et al. [6], in 2012, claimed that Wu et al.'s protocol suffers from lack of anonymity, disclosure of password and imperfect forward secrecy. Hence, they proposed an improved elliptic curve based technique. In 2014, Zhao et al. [7] claimed that Mun et al's scheme is vulnerable to

P. K. Roy · S. Ball · K. Parai (✉)
Siliguri Institute of Technology, Siliguri, India
e-mail: krittibas.sit@gmail.com

P. K. Roy
e-mail: prasanta201284@gmail.com

S. Ball
e-mail: sathiball@gmail.com

© Springer Nature Singapore Pte Ltd. 2019
J. K. Mandal et al. (eds.), *Contemporary Advances in Innovative and Applicable Information Technology*, Advances in Intelligent Systems and Computing 812,
https://doi.org/10.1007/978-981-13-1540-4_19

insider attack and impersonation attack, and does not accomplish user anonymity, user friendliness, proper mutual authentication and login phase verification, and proposed an enhancement. Xie et al. [8] also proposed an anonymous authentication scheme for multiple *MU* environments in 2014. In 2015, Memon et al. [9] proposed an *ECC* based construction where user authentication and shared session key agreement between a mobile client and a base transceiver station are accomplished via mobile client's location-based services. They also found that Zhao et al.'s construction is susceptible to disclosure of mobile client's password and man-in-the-middle attack. Reddy et al. [10], in 2016, argued that Memon et al.'s scheme does not achieve proper mutual authentication, secure password changing phase, protection against impersonation attack and resistance from privileged insider attack, and suggested their construction based on *ECC*. Besides all of these mentioned above, numerous authentication schemes still exist in recent research.

In this article, we have highlighted several limitations of Xie et al.'s protocol [8] and propose an efficient secure anonymous two-factor authentication protocol for roaming services in *GLOMONET*. The proposed protocol is based on *ECC* and constructed using lightweight operations to ensure the resistivity of the said scheme against several well-known security threats with low computation cost.

The rest of this article is summarized as follows. The cryptanalysis of Xie et al.'s construction has been made in Sect. 2. Our proposed scheme along with related issues are demonstrated in Sect. 3. Section 4 demonstrates the performance analysis phase of our protocol. Section 5 presents the formal verification of our protocol using widely accepted AVISPA tool. Finally, our concluding remarks along with future scope of this work have been presented in Sect. 6.

2   Cryptanalysis Against Xie et al.'s Protocol

The cryptanalysis of Xie et al.'s construction [8] is demonstrated below. The notations $ID_X, T_X, T'_X, \Delta t, h(\cdot), G, E_K/D_K,$ and $x_0, y_0, z_0 \in Z_q^*$ represent identity of an entity X, timestamp of X, current timestamp of X, a valid time interval, collision resistant one-way hash function, a point on elliptic curve, encryption/decryption function using key K, and freshly generated random numbers, respectively.

Denial of service: On receipt of *MU*'s login request, *FA* checks whether $T'_{MU} - T_{MU} < \Delta t$ holds. If satisfied, *FA* proceeds. Let us assume, an attacker ($Æ$) holds the login request of *MU* for a period of $\geq \Delta t$ and forwards it to *FA* without tampering the message. *FA* checks the validity of Δt and obviously it will not pass. Hence, the incoming request will be denied immediately though the request sent by a legitimate *MU*. Thus, vulnerable to denial of service.

Key compromised impersonation attack: In this kind of attack if $Æ$ knows the long term secret key of one of the communicating parties, he can form a malicious attack impersonating that party [11]. In Xie et al., if the long term secret key of *FA* (SK_{FH}) is compromised, $Æ$ can easily impersonate *MU*, *FA* and *HA* as follows.

Step 1: Æ intercepts the message $m_1 = \{ID_{HA}, T_{MU}, E_1, X\}$ where, $X = x_0 G$, and computes $Z = z_0.G$. Then, he sends $m_1^* = \{ID_{HA}, T_{MU}, E_1, Z\}$ to FA instead of sending actual m_1. Æ saves m_1 for future reference.

Step 2: FA verifies T_{MU} and generates $Y = y_0.G$. Then, FA computes $V = h\big(SK_{FH}\|m_1^*\|ID_{FA}\|T_{FA}\|ID_{HA}\|Y\big)$ and sends $m_2 = \{m_1^*, ID_{FA}, T_{FA}, V, Y\}$ to HA.

Step 3: Æ again intercepts m_2, saves it for future use, and computes $V^* = h(SK_{FH}\|m_1\|ID_{FA}\|T_{FA}\|ID_{HA}\|Z)$. Æ sends $m_2^* = \{m_1, ID_{FA}, T_{FA}, V^*, Z\}$ to HA.

Step 4: It is obvious that all the verifications done by HA will hold successfully. Hence, HA computes $E_2 = E_{SK_{FH}}[h(SID_{MU}\|DH_{MH}\|T_{MU}\|Z)\|T_{HA}\|X]$ and sends $m_3 = \{E_2, T_{HA}, X\}$ to FA.

Step 5: Æ intercepts m_3, computes $E_2^* = E_{SK_{FH}}[h(SID_{MU}\|DH_{MH}\|T_{MU}\|Z)\|T_{HA}\|Z]$, and sends $m_3^* = \{E_2^*, T_{HA}, Z\}$ to FA.

Step 6: FA decrypts E_2^* using SK_{FH}, and verifies T_{HA} and Z. Then, FA computes $sk_1 = y_0.Z = y_0.z_0.G$ and the session key $SK_1 = h(sk_1\|Z\|Y)$. FA sends $m_4 = \{E_3, Y, Z\}$ to MU where, $E_3 = E_{sk_1}[h(SID_{MU}\|DH_{MH}\|T_{MU}\|Z)\|Z]$.

Step 7: Æ blocks m_4, computes $sk_1 = z_0.Y = y_0.z_0.G$, $SK_1 = h(sk_1\|Z\|Y)$, $sk_2 = z_0.X = x_0.z_0.G$, and $SK_2 = h(sk_2\|X\|Z)$. Then, Æ decrypts E_3 using sk_1 and computes $E_3^* = E_{sk_2}[h(SID_{MU}\|DH_{MH}\|T_{MU}\|Z)\|X]$. Æ sends $m_4^* = \{E_3^*, Z, X\}$ to MU.

Step 8: MU computes $sk_2 = x_0.Z = x_0.z_0.G$, decrypts E_3^* using sk_2 and verifies the decrypted content. It is obvious that the verification process will pass successfully. Then, MU computes the session key $SK_2 = h(sk_2\|X\|Z)$.

Hence, Æ has successfully established the session keys SK_1 and SK_2 between itself and FA, and between itself and MU, respectively. Moreover, Æ may act as a middle man between FA and MU, and create separate sessions with them using session keys SK_1 and SK_2, respectively. Thus, Xie et al.'s protocol is also vulnerable to key compromised impersonation attack and man-in-the-middle attack.

Unverified login phase: To generate a login request, MU inserts his SC into a card reader and inputs its password (PW_{MU}). Unfortunately, SC generates login request m_1 without verifying the legitimacy of MU. Assume, MU mistakenly inputs a wrong password. Still, SC will generate a login request and redirect it to FA without any verification. Thus, unnecessarily increasing the computation cost and may result in extra communication overhead.

Improper mutual authentication: In case of proper mutual authentication, both sender and receiver should verify the authenticity of each other [12]. In Xie et al., after computing SK, FA sends the message $m_4 = \{E_3, Y, X\}$ to MU. Upon receiving m_4, MU firstly verifies the legitimacy of FA and then computes SK. Unfortunately, Xie et al. does not provide any kind of authentication of MU by FA. Hence, there is no way for FA to ensure whether MU has established exactly the same session key as he does. Thus, increasing the chance of malicious attack due to improper mutual authentication.

Excessive computation cost during password changing phase: In Xie et al., each time MU wants to change his password, he needs to go through the verification process with HA and FA. If the protocol was designed in such a way so that MU can freely change his password without involving HA or FA or both, the excessive computation cost due to verification process with HA and FA can be reduced [10].

Clock synchronization problem: The entire protocol is based on the verification of T_{MU}, T_{FA} and T_{HA}. Hence, all the clocks generating these timestamps must be synchronized properly and this may result in extra computation cost.

3 Proposed Protocol

This section demonstrates our proposed protocol with three different phases: registration, login and authentication with session key agreement, and password changing phase. During initialization, HA announces its public key $P_{HA} = S_{HA} \cdot G$ where, S_{HA} is the secret key of HA.

3.1 Registration Phase

To access the ubiquitous services of *GLOMONET*, a new MU has to register his credentials to its HA via a secure channel. First, MU chooses $a \in Z_q^*$ at random and generates $AID_{MU} = h(ID_{MU}\|a)$. Then, MU submits his registration request $\{AID_{MU}, h(\cdot)\}$ to HA. HA computes $AID_{MH} = h(AID_{MU}\|ID_{HA})$ and stores the parameters $\{AID_{MH}, ID_{HA}, h(\cdot), G\}$ on a new SC and assigns it to MU. MU selects PW_{MU}, computes $M_a = h(ID_{MU}\|PW_{MU}) \oplus a$, and inserts M_a into SC. Finally, SC contains the parameters $\{AID_{MH}, ID_{HA}, M_a, h(\cdot), G\}$.

3.2 Login and Authentication with Session Key Agreement Phase

Before establishing SK, both the parties involved in the communication process must mutually authenticate each other via insecure channel as demonstrated in Fig. 1. MU inserts his SC into a card reader and inputs ID_{MU} and PW_{MU}. Then, SC computes $a = M_a \oplus h(ID_{MU}\|PW_{MU})$ and verifies $AID_{MH} \overset{?}{=} h(h(ID_{MU}\|a)\|ID_{HA})$. If successful, SC generates $x \in Z_q^*$ freshly, and computes $X = x \cdot G$ and $XID_{MH} = h(AID_{MH}\|X)$. SC sends the login request $m_1 = \{XID_{MH}, ID_{HA}, X, G, h(\cdot)\}$ to FA. Upon receiving m_1, FA generates $y \in Z_q^*$ freshly, and computes

$Y = y \cdot G$, $K_{FH} = y \cdot P_{HA}$ and $K_{XY} = h(K_{FH}\|X\|Y)$. Then, FA sends the message $m_2 = \{ID_{FA}, XID_{MH}, K_{XY}, X, Y\}$ to HA. After receiving m_2, HA computes $K_{FH} = S_{HA} \cdot Y$, and checks received XID_{MH} and K_{XY} using its own AID_{MH} and K_{FH}, respectively. Then, HA computes $V_1 = h(K_{FH}\|X\|Y)$ and $V_2 = h(AID_{MH}\|X\|Y)$, and sends $m_3 = \{V_1, V_2\}$ to FA. FA verifies V_1 using its own K_{FH} and computes $SK = y \cdot X = x \cdot y \cdot G$. Then, FA sends $m_4 = \{V_2, K_1, Y\}$ to MU where, $K_1 = h(SK\|X\|Y)$. SC verifies V_2 using stored AID_{MH} and received Y. After successful verification, MU computes $SK = x \cdot Y = x \cdot y \cdot G$ and again verifies received K_1. If verified, MU further computes $K_2 = h(SK\|K_1\|X\|Y)$ and sends $m_5 = \{K_2\}$ to FA. On successful verification of received K_2, FA accepts the session key SK. Finally, the session key SK has been securely established between MU and FA.

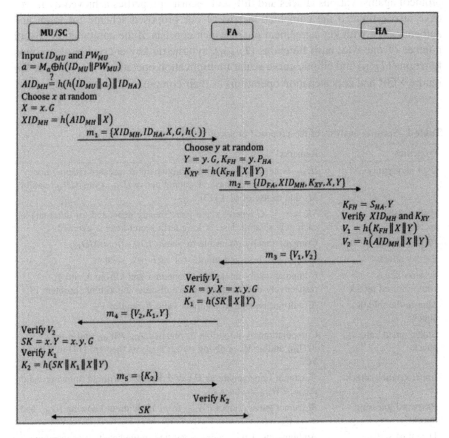

Fig. 1 Login and authentication with session key agreement phase

3.3 Password Changing Phase

In this phase, MU can freely renew his password using its SC and without involving HA and/or FA. First, MU inserts his SC into the card reader and inputs ID_{MU}, PW_{MU}, and the new password PW_{MU}^*. Then, SC computes a and verifies AID_{MH} using ID_{MU} and PW_{MU} as mentioned in Sect. 3.2. If verified, SC computes the new parameter $M_a^* = h(ID_{MU}\|PW_{MU}^*) \oplus a$ and stores it replacing M_a.

4 Performance Analysis

This section analyzes the performance of our construction in terms of security as well as computation cost. Table 1 demonstrates the resistance of our proposed protocol against various attacks and different security properties achieved by it. We have also estimated the computation cost of our proposed scheme for login and authentication with key agreement phase which consists of the computation cost of number of one-way hash functions (T_{HASH}), symmetric key encryption/decryption functions ($T_{E/D}$) and elliptic curve scalar multiplication operations (T_{MUL}) used. We ignore XOR and concatenation operations as their computation costs are negligible.

Table 1 Security analysis of the proposed protocol

Properties	Remarks
User anonymity	ID_{MU} is not disclosed neither anywhere in insecure channel nor during registration phase. Æ cannot derive ID_{MU} from AID_{MU} and/or M_a due to use of $h(\cdot)$ [13].
Perfect forward secrecy	$SK = x \cdot y \cdot G$ where, x and y are freshly generated (or different) in each new session. Use of long term secret keys is avoided.
Insider attack	Computationally infeasible to obtain ID_{MU} from AID_{MH}.
Replay attack	x and y are fresh and unique for each new session.
Session key compromised attack	Computationally infeasible to obtain x and y from X and Y, respectively due to elliptic curve discrete logarithmic problem [9].
Man-in-the-middle attack	Æ will not be able to compute valid V_1 and/or V_2.
Stolen smart card attack	Computationally infeasible to obtain ID_{MU},PW_{MU} and a from AID_{MH} and/or M_a applying power analysis theorem [14] to a stolen SC.
Impersonation attack	Æ cannot impersonate as HA or FA or MU without knowing valid ID_{MU}, PW_{MU}, a and K_{FH}.
Password guessing attack	Æ cannot guess correct PW_{MU} from M_a without knowing ID_{MU} and a.
Denial of service	Authenticity of the sender is verified immediately after receiving each message. No timestamp is used to ensure authenticity of the sender.

Table 2 Computation cost comparison

Protocol	Computational cost	Estimated time (seconds)
Mun et al. [6]	$4T_{MUL} + 2T_{E/D} + 10T_{HASH}$	0.2747
Zhao et al. [7]	$9T_{MUL} + 8T_{E/D} + 16T_{HASH}$	0.645275
Xie et al. [8]	$6T_{MUL} + 7T_{E/D} + 9T_{HASH}$	0.44385
Memon et al. [9]	$4T_{MUL} + 20T_{HASH}$	0.2623
Reddy et al. [10]	$7T_{MUL} + 22T_{HASH}$	0.452525
Our protocol	$6T_{MUL} + 14T_{HASH}$	0.38545

The computation cost of our construction along with Mun et al. [6], Zhao et al. [7], Xie et al. [8], Memon et al. [9] and Reddy et al. [10] are summarized in Table 2 considering the time required for T_{HASH}, $T_{E/D}$ and T_{MUL} as 0.0005 s, 0.0087 s and 0.063075 s, respectively [8]. It can be easily verified that our protocol achieves better computation cost than that of Zhao et al., Xie et al. and Reddy et al. However, it achieves slightly higher computation cost than Mun et al. and Memon et al. but, our protocol achieves more security and design features.

5 Formal Verification Using AVISPA

We have utilized the AVISPA tool [15] to formally verify the safety of our protocol against several security threats under OFMC back-end [16]. The implementation phase of the proposed protocol is done using HLPSL [17] considering the Dolev-Yao threat model [18] to reflect the properties of an insecure channel among *MU*, *FA* and *HA*. A symmetric key is used to implement the secure channel during

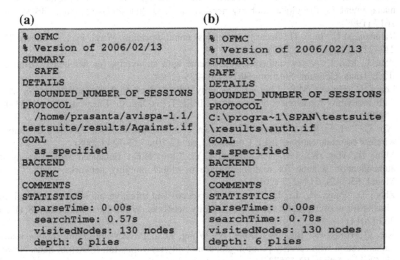

(a)
```
% OFMC
% Version of 2006/02/13
SUMMARY
  SAFE
DETAILS
  BOUNDED_NUMBER_OF_SESSIONS
PROTOCOL
  /home/prasanta/avispa-1.1/
testsuite/results/Against.if
GOAL
  as_specified
BACKEND
  OFMC
COMMENTS
STATISTICS
  parseTime: 0.00s
  searchTime: 0.57s
  visitedNodes: 130 nodes
  depth: 6 plies
```

(b)
```
% OFMC
% Version of 2006/02/13
SUMMARY
  SAFE
DETAILS
  BOUNDED_NUMBER_OF_SESSIONS
PROTOCOL
  C:\progra~1\SPAN\testsuite
  \results\auth.if
GOAL
  as_specified
BACKEND
  OFMC
COMMENTS
STATISTICS
  parseTime: 0.00s
  searchTime: 0.78s
  visitedNodes: 130 nodes
  depth: 6 plies
```

Fig. 2 HLPSL implementation result: (**a**) Our proposed scheme. (**b**) Reddy et al. [10]

registration phase of our protocol. Two insecure channels are considered: between *MU* and *FA*, and between *FA* and *HA*. We have fixed six secrecy goals and two authentication goals namely, secrecy of ID_{MU}, secrecy of PW_{MU}, secrecy of a, secrecy of x, secrecy of y, secrecy of S_{HA}, authentication of *MU* and authentication of *FA*. Figure 2 shows the HLPSL implementation result of our scheme along with Reddy et al.'s scheme [10]. It can be easily verified that our protocol remains 'SAFE' under Dolev-Yao channel and achieves all the specified secrecy and authentication goals with a search time of 0.57 s.

6 Conclusion

This article highlights several limitations of Xie et al.'s construction and demonstrates a secure two-factor anonymous authentication protocol for roaming services in *GLOMONET*. The security analysis and formal verification phases of the proposed scheme prove its strength against several security threats. Also, the computation cost of our construction is better than Xie et al.'s scheme and several existing schemes. Hence, our protocol is more efficient than Xie et al.'s scheme and suitable for low power practical applications of *GLOMONET*. Our future work includes an extension of this protocol towards three-factor authentication scheme satisfying all kind of security and design goals.

References

1. Suzukiz, S., Nakada, K.: An authentication technique based on distributed security management for the global mobility network. IEEE J. Sel. Areas Commun. **15**(8), 1608–1617 (1997)
2. Rahman, M.G., Imai, H.: Security in wireless communication. Wirel. Pers. Commun. **22**(2), 213–228 (2002)
3. Zhu, J., Ma, J.: A new authentication scheme with anonymity for wireless environments. IEEE Trans. Consum. Electron. **50**(1), 231–235 (2004)
4. Lee, C.C., Hwang, M.S., Liao, I.E.: Security enhancement on a new authentication scheme with anonymity for wireless environments. IEEE Trans. Industr. Electron. **53**(5), 1683–1687 (2006)
5. Wu, C.C., Lee, W.B., Tsaur, W.J.: A secure authentication scheme with anonymity for wireless communications. IEEE Commun. Lett. **12**(10), 722–723 (2008)
6. Mun, H., Han, K., Lee, Y.S., Yeun, C.Y., Choi, H.H.: Enhanced secure anonymous authentication scheme for roaming service in global mobility networks. Math. Comput. Model. **55**(1–2), 214–222 (2012)
7. Zhao, D., Peng, H., Li, L., Yang, Y.: A secure and effective anonymous authentication scheme for roaming service in global mobility networks. Wirel. Pers. Commun. **78**(1), 247–269 (2014)
8. Xie, Q., Hong, D., Bao, M., Dong, N., Wong, D.S.: Privacy-Preserving mobile roaming authentication with security proof in global mobility networks. Int. J. Distrib. Sens. Netw. **10** (5) (2014). Article ID 325734

9. Memon, I., Hussain, I., Akhtar, R., Chen, G.: Enhanced privacy and authentication: an efficient and secure anonymous communication for location based service using asymmetric cryptography scheme. Wirel. Pers. Commun. **84**(2), 1487–1508 (2015)

10. Reddy, A.G., Das, A.K., Yeen, E.J., Yoo, K.Y.: A secure anonymous authentication protocol for mobile services on elliptic curve cryptography. IEEE Access **4**, 4394–4407 (2016)

11. Lu, Y., Li, L., Yang, Y.: Robust and efficient authentication scheme for session initiation protocol. Math. Probl. Eng. (2015). Article ID 894549

12. Boyd, C., Mathuria, A.: Protocols for authentication and key establishment, Springer Science and Business Media (2013)

13. Rogaway, P., Shrimpton, T.: Cryptographic hash function basic: definitions, implications, and separations for preimage resistance, second-preimage resistance, and collision resistance. In: Proceedings of International Workshop on Fast Software Encryption (2004), pp. 371–388

14. Kocher, P., Jaffe, J., Jun, B.: Differential power analysis. In: Proceedings of Advances in Cryptololgy (CRYPTO'99), vol. 1666 (1999), pp. 388–397

15. Armando, A.: The AVISPA Tool for the Automated Validation of Internet Security Protocol and Applications. Computer Aided Verifcation, Springer, Berlin, Germany, (July, 2005), pp. 281–285

16. Basin, D., Modersheim, S., Vigano, L.: OFMC: a symbolic model checker for security protocol. Int. J. Inf. Secur. **4**(3), 181–208 (2005)

17. Oheimb, D.V.: The high-level protocol specification language HLPSL developed in the EU project AVISPA. In: Proceedings of APPSEM Workshop (September 2005), pp. 1–17

18. Dolev, D., Yao, A.C.: On the security of public key protocols. IEEE Trans. Inf. Theory **29**(2), 198–208 (1983)

Comparative Performance Analysis of DTN Routing Protocols in Multiple Post-disaster Situations

Amit Kr. Gupta, Jyotsna Kumar Mandal and Indrajit Bhattacharya

1 Introduction

The huge loss of life and property associated with recurring disasters all over the world require immediate involvement of the active research community for devising a practical solution. The disaster being referred here is any large scale natural or man-made disaster that has huge impact on masses and property. A severe flood, earthquake, train-accident, etc., are examples of such disasters. A post-disaster situation refers to the havoc condition of the disaster affected area just after the disaster has struck. The post-disaster situation presents a condition where a large mass of people gets affected that includes loss of life and property. It is a common observation in a post-disaster situation that the existing communication infrastructure of the area also gets damaged. The scale of destruction depends on the severity of the disaster and proneness of the disaster affected area. There are certain identifiable areas on land that are more vulnerable to huge impact of disasters because of their proneness. A feasible and efficient post-disaster tool for analyzing the situation and presenting a resource management strategy is need of the time that will help to reduce the after effects of such disasters.

It has been found in the post analysis of major disasters all over the world that most of the casualties occur during the initial early period (first few hours after

A. Kr. Gupta (✉) · J. K. Mandal
Department of Computer Science and Engineering,
University of Kalyani, Kalyani, India
e-mail: amitgupta.kgec@gmail.com

J. K. Mandal
e-mail: jkm.cse@gmail.com

I. Bhattacharya
Department of Computer Applications, Kalyani Government
Engineering College, Kalyani, West Bengal, India
e-mail: indra51276@gmail.com

© Springer Nature Singapore Pte Ltd. 2019
J. K. Mandal et al. (eds.), *Contemporary Advances in Innovative and Applicable Information Technology*, Advances in Intelligent Systems and Computing 812,
https://doi.org/10.1007/978-981-13-1540-4_20

disaster strikes) and that too because of absence of information about the struck life in interior locations of the disaster affected area. The current relief mechanism in most of the developing nations depends on direct involvement of human beings. The entire relief work is handicapped by whatever information gets collected by the relief workers. The emphasis of word handicapped is of utmost importance here as human beings have limited capabilities and resources to reach to the remote locations of the disaster affected areas. In the current time of computer advancement enabling humans to reach moon and other heavenly bodies, it seems quite practicable to utilize the capabilities of Information and Communication Technology (ICT in short) for such relief work. As the existing communication infrastructure collapses after a major disaster, ICT can be efficiently utilized in creation of a fast and easily deployable, temporarily built delay tolerant network (DTN) [1] as the basis of communication infrastructure in a post-disaster environment. Some works on the post-disaster network architectures utilizing the features of a DTN are already existent like the four-tier network architecture as presented in [2]. Human beings, with their limited capabilities, are restricted to work at tier-1 of the architecture and other tasks are carried over by various ICT devices like smart phones, UAVs, Vehicles, towers, etc. The data collection in the said architecture is automated by using handheld devices and suitable software protocols. For such data collection, some routing protocols must be present that define the way data is to be collected. Such routing protocols have also been proposed in the literature. The specific DTN type architecture necessitates DTN routing protocols to work over them. Some of the DTN routing protocols that work effectively in a post-disaster scenario are Epidemic [3], Prophet [4], Spray and Wait [5], MaxProp [6] and DirMove [2]. The performance of DTN routing protocols heavily rely on the underlying mobility models that define the pattern of movement of the relief workers and other stakeholders in a post-disaster environment who work in a coordinated manner and they often move in some predefined effective way. The current paper presents a comparative performance analysis of these DTN protocols over three different types of post-disaster environments or mobility models, which are nothing but emulation of the real-world scenario of movement patterns as happens during a post-disaster situation. The purpose of the current work is to propose a suitable DTN routing protocol that can be effectively implemented during need arising out of occurrence of any large scale disaster.

Rest of this paper is presented as explained. Section 2 in the paper presents a literature survey of DTN routing protocols, post-disaster Mobility models and related type of performance analysis done in the past. Section 3 contains detailed description of the three different types of post-disaster environment (mobility models) and the performance parameters that have been utilized in our survey. The comparison analysis results thus obtained have been shown in Sect. 4 of the paper. Section 5 presents conclusion to the undertaken analysis work. Future scope of the current work is proposed in Sect. 6.

2 Related Work

Data communication in a DTN environment has unique challenges of its own that relate to specific features of a DTN like intermittent connectivity, uncertain and long delays, network partitions, and so on. Overcoming these challenges, some routing protocols utilizing a DTN have been presented in literature that includes Epidemic, the Prophet, Spray and Wait strategy, MaxProp and DirMove. Each of these routing protocols is described in brief in the following part.

In Epidemic routing, all messages from a node get copied and are transmitted to every other nodes arriving within its transmission range as the nodes continuously move in the network. Epidemic provides measurable good delivery ratio incurring a small delivery delay but it suffers from huge buffer occupancy, large number of packet drops and very high energy demands that render it unsuitable in a post-disaster situation.

The Prophet routing protocol keeps a check on the copies of a message that get distributed in the network. On two different nodes meeting each other, they exchange their delivery predictability information and the index vector as in Epidemic with each other. This information is used to calculate probability of a node's ability to deliver a message to its destination. In the Prophet routing method, a message gets delivered to another node only if the calculated delivery pre- dictability concerning the destination node is higher at the other node. Prophet's working depends entirely on the information of the past performances. Absence and lack of current information about the direction of movement of the nodes renders it go haywire, resulting in increased latency. Lack of acknowledgement incurs huge buffer occupancy in Prophet.

In Spray and Wait method, the entire work is divided between two different phases. The first one is the Spray phase where a calculated and limited number of copies of any message get spread in the network by the source node including some other nodes which later obtain a message replica. In the next phase called the Wait phase, nodes that have spread copies of their messages wait until the destination is not accessed by any node carrying a replica of the message (from the spraying phase). After this, each of the nodes that carry a message's copy try in their turn to deliver their copies to the destination involving a direct transmission. Spray and wait suffers long latency and inefficient use of bandwidth. Requirement of high mobility for deepest node to reach destination directly does not seem realistic in a DTN environment for a post-disaster situation.

The MaxProp protocol utilizes a scheme to prioritize scheduling of packets that are transmitted to some other nodes as also it prioritizes the schedule of those packets that must be removed from buffer. MaxProp, similar to the Prophet, con- siders the past performances and hence renders itself unsuitable in a post-disaster like situation.

DirMove is based on forwarding decisions by a node on the basis of the direction of movement of receiving nodes. The DTN routing protocol keeps a track of the recent direction of movement of moving nodes. It measures each node's consecutive distances from the destination node at two different but close enough instants. If a node is moving away from the destination (in opposite direction) between two consecutive time instants, then that node seems to be very unlikely to carry the messages to the destination. The fittest node among all neighboring nodes of a node gets chosen to be the next hop for that instance. Selection of fittest node is done by utilizing calculations involving parameters such as past delivery history and delivery delay, current movement direction and a node's recent proximity to destination. Certain issues that relate to routing in a DTN such as a node's fitness for successful message delivery, good buffer management, smaller packet drops and issues relating to node energy are appropriately considered in the DirMove.

A number of survey works have also been carried out in the past that compare various DTN routing protocols for their performance analysis. The work done in [7] shows a comparative study of various DTN routing strategies for their performance over a cluster-based mobility model. In this work, authors have found out that MaxProp and Prophet routing protocols are superior in their performance to others when a cluster mobility model is in consideration. This work has limited scope of application as disaster relief work may not always be limited to cluster-based mobility, where other models may also be followed. The authors in [8] have suggested some of the DTN routing protocols that are suitable to work in a post-disaster scenario but no standardized comparison has been made to suggest better or good protocols. The DTN routing comparison works done in [9, 10] are based on a single mobility model and different routing methods have been evaluated over it. The concept of performance comparison over multiple mobility patterns is very novel and presents a scope of real-life implementation in case of any large scale disaster.

Mobility models are divided into broad categories—namely Entity-Based model and the other one as the Group-based mobility model [11]. Nodes move individually with no influence by other nodes in an entity-based model, whereas in Group-based model the nodes' movement within groups is affected by other member nodes. In the Random Waypoint [11] model, which is an Entity-Based mobility model, mobile nodes select destination points randomly and travel there with constant speed and some pauses at destinations. Random Walk [11] is again an Entity-Based mobility model which is similar to a Random Waypoint model but having zero pause time. The Shortest Path Map Based mobility model [11] is an Entity-based and map based model which exploits algorithms such as the Dijkstra's to calculate shortest paths between any two points on the map. Working day mobility model [11] is a Group-based model that models an overall result of various sub-models of node mobility during an entire day. It considers day to day common activities of different types of people. Cluster Mobility Model [11, 12] is a group-based model that divides the entire network in a certain number of clusters. Nodes that carry data from one cluster to another are Carrier nodes. The other nodes present in each cluster are known as internal nodes. An internal node's movement gets defined around a Cluster Center. Cluster Mobility Model is best suited as a group-based mobility model to emulate a post-disaster scenario.

Uddin et al. [13] have proposed a post-disaster mobility model for a DTN that helps in providing communication in such contexts where it is infeasible and difficult to think of a guaranteed end-to-end connectivity. The mobility model proposed by them utilizes various actors in a post-disaster including relief workers of various categories, transportation network, population movement and relief vehicle movement, etc.

All of the above-mentioned comparison works present performance analysis over a single environment and are not exhaustive in nature to suggest the best generalized DTN protocol to work efficiently over multiple situations. The current work presents a comparative performance analysis of some of the best performing DTN protocols over three different types of post-disaster environments and for a multiple number of performance parameters. This helps in finding a DTN routing method that is generally suitable for its application in literally any type of post-disaster environment. The considered disaster mobility models and measuring performance parameters have been discussed next in Sect. 3.

3 Post-disaster Environments and Performance Indicators

A post-disaster environment mentioned here generally refers to the early recovery stages just after the disaster has struck. During recovery, humanitarian and relief providing teams arrive on the affected area with their personnel and get to work following some standard protocols. The relief workers and other stakeholders in a post-disaster environment work in a coordinated manner and they often move in some predefined movement patterns that have been proven to be effective. These patterns are termed as Mobility Models. In the current work, three different post-disaster mobility models have been studied that are discussed in the following sections.

3.1 Post-disaster Mobility Models

The current work considers multiple post-disaster environments to analyze DTN protocols for their generic utilization. Simulation methodology has been chosen to analyze performances of existing DTN routing protocols in a post-disaster scenario using ONE (Opportunistic Networks Environment) simulator. The environments have been created as mobility models in the ONE simulator by varying simulation parameters. The Cluster Mobility model, as discussed in Sect. 2 has been modeled in the ONE simulator with the settings as defined in Table 1.

The other mobility model that has been simulated for comparisons in this work is the event- and role-based post-disaster mobility model using the following settings in the ONE Simulator as presented in Table 2.

Table 1 Parameter values for the cluster mobility model

Parameter	Value
Complete simulation time	43,200 s = 12 h
Update interval	1 s
Number of nodes	120 [(25 nodes × 4 clusters) + 20 carrier_nodes]
Buffer size of each node	500 MB
Speed of cluster nodes	0.5–1.5 mps = 1.8–5.4 kmph
Speed of carrier nodes	5–15 mps = 18–54 kmph
Scan interval of the nodes	0 s
Wait time of cluster	0–2 min
Wait time of carrier nodes	0–10 min
Message time to live	240 min = 4 h
Message creation time interval	25–120 s
Each message size	50 kB–1 MB
Simulation area	12 km^2 (4 km × 3 km)

Table 2 Parameter values for the post disaster mobility model

Parameter	Value
No. of civilians	100
No. of police/fire fighters	50
No. of ambulances	50
No. of hospitals	4
No. of events	4
Simulation area size	3000 m * m
Simulation time	2000 s
No. of nodes	204
Interface	Bluetooth
Range of transmission	250 m
Time to live	50 min
Message interval rate	At every 10–20 s

A third mobility model emulating the conditions in a post-disaster scenario that we have selected in this paper is the Random waypoint mobility model with the below-mentioned properties defined in Table 3.

These mobility models define the basis for performing simulations in the ONE simulator where any suitable DTN routing protocol can be implemented to execute. After carrying out these simulations, the results are compiled using some standard performance metrics which are discussed next.

Parameters	Values
Affected area	4.5 km × 3.4 km
No. of nodes	50
Transmission range (in m)	250 m
Transmission speed (in Mbps)	2 Mbps
Node speed (in km/h)	10–50 km/h
Time to live of message (in min)	60 min
Buffer size	Infinite
Movement model	Random waypoint
Simulation time (in min)	320 min
Message creation rate	1 per 10–15 s
Message size	500 kB

Table 3 Parameter values for the random waypoint mobility model

3.2 Performance Metrics

The commonly used metrics that are generally selected to evaluate an analysis of the performance of various routing protocols in a disaster scenario have been discussed here. These metrics are the standard performance indicators for analyzing DTN routing protocols. These metrics include:

- Delivery Ratio: It provides a measure of the ratio of number of successfully delivered messages to the total number of messages created.
- Average Delivery Latency/Delay: Latency, in general, is the measure of time gap between creation of a message at the source node followed by its delivery at the destination. Average delay gives the sum of latencies of all messages that were delivered successfully.
- Overhead Ratio: It provides a measure of the count of redundant messages relayed to successfully transmit one message to the destination. It is explained using the following equation:

$$\text{Overhead Ratio} = (R - D)/D \qquad (1)$$

where R shows the number of messages that are relayed and D gives the count of the number of messages delivered to the destination successfully.

4 Results, Analysis, and Discussion

The simulations in this work have been performed over the ONE (Opportunistic Network Environment) Simulator, which has established itself to be industry standard for researches in the field of DTN. Initially, a mobility model is configured

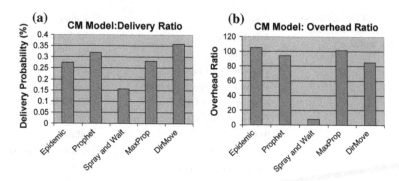

Fig. 1 Performance depiction in the cluster mobility model: **a** message delivery ratio, **b** message overhead ratio

in the ONE and then the various routing protocols are executed to measure their performance. At first, the Cluster Mobility Model has been configured and then the routing methods were executed and their delivery and overhead ratios were measured. Similarly, the simulations have been performed with the Post-disaster and the Random waypoint mobility models. The simulation results so obtained have been presented using graphs in Figs. 1, 2 and 3. The performance comparison of the

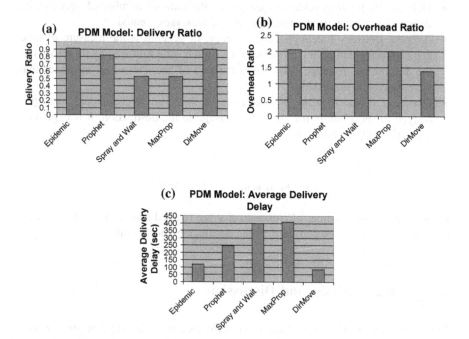

Fig. 2 Performance depiction in the post disaster mobility model: **a** message delivery ratio, **b** message overhead ratio, **c** average delivery delay

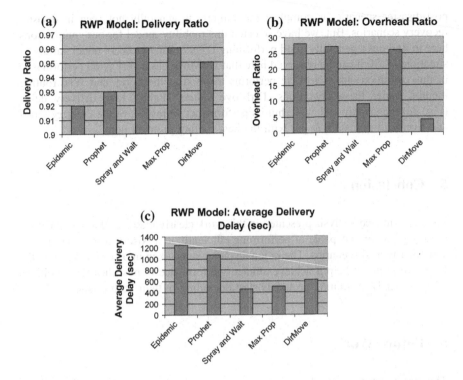

Fig. 3 Performance depiction in the random waypoint mobility model: **a** message delivery ratio, **b** message overhead ratio, **c** average delivery delay

selected DTN routing protocols over the cluster mobility model has been presented in Fig. 1. It can be observed here that DirMove performs the best under a cluster mobility that is good depiction of a post-disaster environment. Though Spray and Wait has least overhead ratio (Fig. 1b), but it suffers heavily in respect of its delivery ratio. DirMove has best delivery ratio under this condition and has a manageable overhead ratio. This is because of the fact that routing decision in DirMove considers a multiple number of parameters that include current parameters as well as parameters relating to past performances.

Graphs in Fig. 2 show a similar trend where DirMove outperforms other routing protocols in the Post-disaster Mobility Model as well. Figure 2a clearly shows that DirMove provides best delivery with least latency (Fig. 2b) and smallest overhead (Fig. 2c). This is attributable to the smart and better routing strategy adopted in the DirMove routing protocol. Consideration of current situational information along with historical data provides a benefit to DirMove in the post-disaster scenario so that it provides good delivery with smallest delay. So, DirMove performs best under the Post-disaster Mobility Model also.

The graphs in Fig. 3 show performance of the selected routing protocols under the Random Waypoint Mobility Model which is not a complete emulation of a post-disaster scenario as complete randomness is generally not seen in disaster recovery scenarios. But, we have selected this mobility model for our comparisons here to find out a generally good performing DTN routing protocol under variety of situations. Still, it can be observed here that DirMove has a very good performance in this condition also. DirMove performs as good as the Spray and Wait routing method under these conditions. DirMove suffers slightly long delivery delay (Fig. 3c) than Spray and Wait, but its performance in respect of the delivery ratios and overhead ratios are at par with the best routing protocol here.

5 Conclusion

The performance analysis presented in this work clearly indicates that the DirMove routing protocol is capable of performing efficiently in a post-disaster environment and other related scenarios. The rationale parameters taken into consideration in the DirMove protocol help it achieve efficient performance in post-disaster conditions where other DTN routing protocols start degrading in their performances.

6 Future Work

The current work focused on comparing unicast DTN routing protocols and proposed a generalized protocol to suit multiple environments. Similarly, the current work will be extended in the future to test performance of multicast DTN routing protocols and will try to propose a general best solution in that case.

References

1. Cerf, V., Burleigh, S., Hooke, A., Torgerson, L., Durst, R., Scott, K., Fall, K., Weiss, H.: Delay-tolerant networking architecture. Internet RFC 4838 (April 2007)
2. Gupta, A.K., Bhattacharya, I., Banerjee, P.S., Mandal, J.K., Mukherjee, A.: DirMove: direction of movement based routing in DTN architecture for post-disaster scenario. In: Wireless Networks, vol. 22, pp. 723–740. Springer US (June 2015)
3. Vahdat, A., Becker, D.: Epidemic Routing for Partially Connected Ad hoc Network. Duke University Technical Report Cs-200006 (April 2000)
4. Lindgren, A., Doria, A., Schelen, O.: Probabilistic Routing in Intermittently Connected Networks, vol. 3126, pp. 239–254 (2004)
5. Spyropoulos, T., Psounis, K., Raghavendra, C.S.: Spray and wait: an efficient routing scheme for intermittently connected mobile networks. In: Proceedings of ACM SIGCOMM Workshop Delay-Tolerant Networking, pp. 252–259 (2005)
6. Burgess, J., Gallagher, B., Jensen, D., Levine, B.N.: MaxProp: routing for vehicle-based disruption-tolerant networks. In: Proceedings of IEEE INFOCOM. 6. https://doi.org/10.1109/infocom.2006.228

7. Saha, S., Sheldekar A., Joseph, C.R., Mukherjee, A., Nandi, S.: Post disaster management using delay tolerant network. In: Özcan A., Zizka J., Nagamalai D. (eds) Recent Trends in Wireless and Mobile Networks. Communications in Computer and Information Science, vol. 162. Springer, Berlin (2011)
8. Jones, E.P., Ward, P.A.: Routing strategies for delay-tolerant networks. In: Submitted to ACM CCR (2006)
9. Agussalim, T.M.: Comparison of DTN routing protocols in realistic scenario. In: International Conference on Intelligent Networking and Collaborative Systems, Salerno, pp. 400–405 (2014). https://doi.org/10.1109/incos.2014.80
10. Alaoui, E.A., Agoujil, S., Hajar, M., Qaraai, Y.: The performance of DTN routing protocols: a comparative study. In: WSEAS Transactions on Communications, vol. 14, pp. 121–130 (2015)
11. Camp, T., Boleng, J., Davies, V.: A survey of mobility models for ad hoc network research. In: Wireless Communication and Mobile Computing (WCMC) (Special issue on mobile Ad hoc networking: research, trends and applications) 2(5), 483–502 (2002)
12. Romoozi, M., Babaei, H., Fathy, M.: A cluster-based mobility model for intelligent nodes at proceeding ICCSA 2009. In: Proceedings of the International Conference on Computational Science and Its Applications: Part I (2009)
13. Uddin, Y.S., Nicol, D.M.: A post-disaster mobility model for delay tolerant networking. In: Proceedings of the 2009 Winter Simulation Conference (2009)

Unique Approaches of Process Modeling for First-Order- and Second-Order System with Examples

Diptarup Bandyopadhyay

1 Introduction

In all the process industries it is very important to analyze the process behavior in the presence of different disturbances. Mathematical modeling is a tool to represent a physical process mathematically. The transfer function of a process allows defining a relation between input and output variables. The conventional approach was to apply the conservation of energy, mass or momentum to define a relationship between input and output state variables. For first-order system the approach is quite easier, whereas the same approach is bit complicated for higher order system. Aim is to identify the each process blocks with electrical components analogy and replace the physical process with electrical circuits. By the help of network analysis the transfer function of the system can be calculated in simpler manner.

2 Theoretical Background

In mostly all the physical process, there is a flow of energy or mass from or to the system. So, when there is a flow of energy or mass and as a result of that some quantity of this parameter is accumulated in the system, it can be compared with electrical resistance and capacitance as below:

A. **Process Resistance**

The Process Resistance [1, 4] will be discussed in two cases.

Mass flow: To understand the process resistance, the flow of liquid through a pipe is considered. When the liquid (incompressible and viscous) is flowing

D. Bandyopadhyay (✉)
Kingston Polytechnic College, Barasat, India
e-mail: diptarupbandyopadhyay@gmail.com

© Springer Nature Singapore Pte Ltd. 2019
J. K. Mandal et al. (eds.), *Contemporary Advances in Innovative and Applicable Information Technology*, Advances in Intelligent Systems and Computing 812,
https://doi.org/10.1007/978-981-13-1540-4_21

Fig. 1 Physical diagram of a
liquid flowing through
capillary

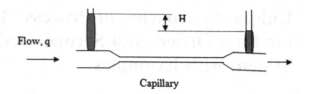

through a pipe, it experiences a shear stress, which results a velocity gradient
between upstream and downstream side. For a capillary tube (as shown in the
Fig. 1), flow rate of liquid through capillary is proportional to this pressure
difference.

As the flow through the capillary is laminar, so here flow of liquid (q) is
analogous to flow of electron (i.e., current) and the head press difference (H) is
analogous to potential difference between two points in upstream and down-
streamside. So here, $H \propto q$ [As Head pressure increases the flow rate will also
increase linearly]

$$i.e. \, H = R \times q \, [R = \text{Flow resistance contributed due to viscous drag}] \quad (1)$$

Equation 1 can be represented by the block diagram of Fig. 2.

In the equivalent electrical circuit is shown in Fig. 3, where $[V_1 - V_2]$ is
equivalent to $[H]$; current $[i]$ is equivalent to flow rate $[q]$ and in both case the
resistance is indicated by $[R]$.

The capillary action is considered where the liquid flows through a restriction.
Energy flow: For any energy balance mechanism, the energy flows from higher
energy level to lower energy level to reach the equilibrium state. The difference
of energy level is analogous to the electrical potential difference. The flow of
energy is analogous to the electrical current. This can be said that the flow of
energy is proportional to the difference of energy level. Again for a particular
energy difference, the amount of energy flow depends on different factor like

Fig. 2 Block diagram

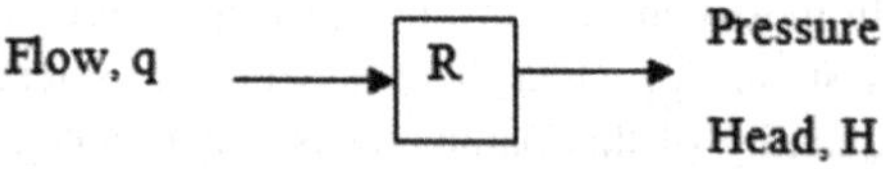

Flow, q $\longrightarrow$ R $\longrightarrow$ Pressure
Head, H

Fig. 3 Equivalent electrical
circuit

R $\qquad V_1 > V_2$

$V_1 \qquad \downarrow \qquad V_2$

i

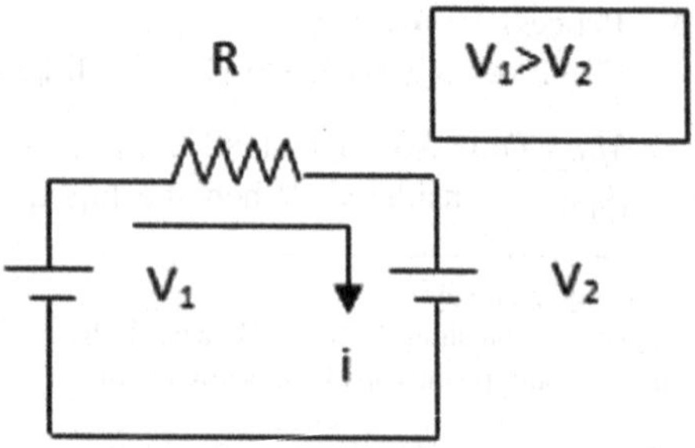

insulation of the medium. This can be treated as energy flow resistance. Hence the value of flow resistance, R will be

$$R = [\text{Difference of energy level}, (E_1 - E_2)]/[\text{Energy flow rate}, (Q)] \qquad (2)$$

B Process Capacitance

Most of the processes are having some storage capability or capacitance. This capacitance of the process helps to store the mass or energy. The Process Capacitance [1, 4] can also be explained in two cases.

Mass storage: The concept is used to store the mass or material of solids, liquids or gases. As example: consider the tank-level system. Here the capacitance or the storage capacity of the tank depend upon the cross-sectional area of the tank for a particular length.

A purely capacitive system is the tank-level system, which have only inflow connection (as shown in Fig. 4). Here the rise of liquid level is proportional to the inlet flow rate and inversely proportional to the process capacitance. As per electrical analogy, if liquid level (H) is compared with rise in voltage in the capacitor during charging (V_C) and liquid inflow rate (q) is compared with charging current (i) for capacitor over a time period "t", then the rise in liquid level will be H = [total volume of water flowing inside the tank in time, t]/[area of the tank, A]

$$\text{i.e.} \, H = [q \times t]/[A] \qquad (3)$$

Equation (3) is equivalent to the voltage across capacitor

$$
\begin{aligned}
V_C &= [\text{current} \times \text{time}][\text{Capacitance}] \\
&= [i \times t]/[C]
\end{aligned}
\qquad (4)
$$

Comparing Eqs. (3) and (4), it can be concluded that process capacitance (i.e. area, A) is equivalent to the electrical capacitance [C].

Energy storage: The concept of energy storage in the most of the process deals with storage of thermal or chemical energy. The thermal capacitance of any material can be calculated by the product of the mass of the body [M] with the specific heat of the material (s) by which the body is

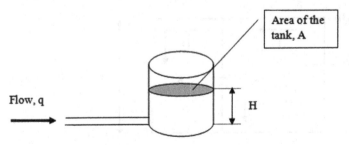

Fig. 4 Capacitive tank-level system

made, per unit degree temperature rise. This thermal capacitance is directly analogous to electrical capacitance. The unit of thermal capacitance is Cal/°C and BTU/°F.

3 Case Study to Calculate Transfer Function with Examples

A. **Single Tank-Level system—Time delay type**:

The Tank-Level system [2–4] is shown here in Fig. 5a. It has one inlet flow connection and one outlet flow connection. The outlet flow rate is the function of liquid level inside the tank. So if level increases, the outlet flow will also increase. For a particular value of inlet flow rate, the tank level reaches a steady state value after certain time. But if there is a sudden load change due to variation of inlet flow rate, the process will indicate that change in terms of next steady state value after certain time lag. So this system is indicated as time delay type or time lag type.

Assume, the volumetric flow rate (volume/time) of liquid inlet is $= F_i$, which is treated as manipulated variable and the volumetric flow rate of liquid outlet is $= F_o$.

The outlet flow resistance is $= R$.

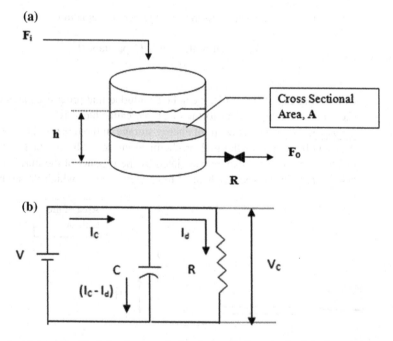

Fig. 5 **a** Schematic of tank-level system, **b** electrical equivalent circuit of single tank-level system

In single tank-level system (from Fig. 5a), it is seen that the tank-level system can be compared with a capacitor charging circuit. The storage capacity of the tank (indicated by the area of the tank, A) is analogous to electrical capacitance (C). The liquid level inside the tank (h) is compared with the voltage across capacitor (V_C). The inlet flow rate of liquid (F_I) is compared with capacitor charging current (I_C). As the liquid outlet flow rate (F_o) linearly varies with liquid level, it can be compared with capacitor leakage current (I_d) through a resistive path R (here it is flow resistance). In steady state condition, the liquid flow rate F_I is constant. So for steady state, the charging current may be considered due to DC supply voltage "V". The equivalent electrical circuit is shown in the Fig. 5b.

Here, the transfer function has to be calculated; i.e.,

$$G(S) = \frac{V_C(S)}{I_C(S)} = ?$$

The capacitor reactance $= \frac{1}{sC}$ which is parallel to resistance R.
Hence equivalent resistance is,

$$R_{eq} = \frac{1}{sC} || R = \frac{R}{1 + sCR} \tag{5a}$$
$$\therefore V_C = I_C \times R_{eq}$$

Hence, $G(S) = \frac{V_C(S)}{I_C(S)} = R_{eq}$
Substituting the value of R_{eq} from Eq. 5a

$$\boxed{\{G(S) = \tfrac{R}{1 + sCR}\}} \tag{5b}$$

Now substituting the value of electrical capacitance "C" by process capacitance "A",

$$\boxed{\{G(S) = \tfrac{R}{1 + ARs}\}} \tag{5c}$$

Hence it is the transfer function single tank-level system.

B. **Non interactive Tank-Level System** (Fig. 6):
Here the process capacity of two tanks are identified by two electrical capacitor "C_1" and "C_2" respectively. It can be explained that the capacitor "C_1" is being charged by the charging current "I_C" (which is equivalent to inlet flow rate "F_i"). Again the capacitor "C_2" is charged by the leakage current "I_d" from the capacitor "C_1" (which is equivalent to outlet flow rate from first tank, i.e. F'). The magnitude of the charging current "I_d" for capacitor "C_2" does not vary with the voltage level at the capacitor "C_2", which is indicated by "V_{C_2}" (as the two systems are non-interacting). Hence the equivalent electrical circuit is

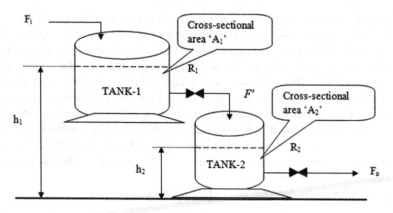

Fig. 6 Schematic for two non-interacting tank-level system

shown in Fig. 7. Consider the leakage path resistance of first capacitor is "R_1" and same for second capacitor is "R_2". The two port network is advisable as the second capacitor is charged via constant current source "I_d". Here (Fig. 7)

$$G(S) = \frac{V_{C_2}(S)}{I_C(S)} = ?$$

For capacitor C_1; Transfer function

$$G(S) = \frac{V_{C_1}(S)}{I_C(S)} = \frac{R}{(1+sC_1R_1)} \tag{6}$$

Again,

$$I_d = \frac{V_{C_1}}{R_1} \tag{7}$$

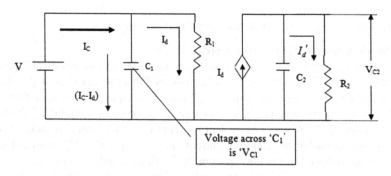

Fig. 7 The electrical equivalent circuit for two non-interacting tank-level system

Similarly for capacitor C_2,

$$\frac{V_{C_2}(S)}{I_D(S)} = \frac{R_2}{(1 + sC_2R_2)} \tag{8}$$

Now from Eq. 7

$$V_{C_2}(S) = I_D(S) \times \frac{R_2}{(1 + sC_2R_2)} = \frac{V_{C_1}}{R_1} \times \frac{R_2}{(1 + sC_2R_2)}$$

$$= \frac{I_C(S)R_1}{R_1(1 + sC_1R_1)} \times \frac{R_2}{(1 + sC_2R_2)} \qquad \text{[From Eq. 6]}$$

$$\therefore \boxed{\left\{ \frac{V_{C_2}(S)}{I_C(S)} = \frac{R_2}{(1 + sC_1R_1)(1 + sC_2R_2)} \right\}} \tag{9}$$

4 Conclusions

This approach will help to deduce the transfer function of any higher order system in a conducive manner.

References

1. Patranabis, D.: Principle of Process Control. TMH publication
2. Stephanopoulos, G.: Chemical Process Control—An Introduction to Theory and Practice
3. Valkenburg, V.: Circuit Theory Foundation
4. Instrumentation Engineers Handbook. BG Liptak

Part VI
Social Network

Study of Social Media Activity of Local Traffic Police Department: Their Posting Nature, Interaction, and Reviews of the Public

Mohammed Allama Hossain, Rashika Daga, Saptarsi Goswami
and Satyajit Chakrabarti

1 Introduction

According to the report generated in 2017 by statistia.com, social media users comprised of 71% of the overall internet users and its estimated that the number of social media users will reach 2.77 billion (bn) from 2.46 bn users [1]. Internet penetration, ease of use and increase worldwide usage of smartphones has been responsible for this virtual connectivity among people. Online connectivity and user engagement provides opportunities to extract meaningful information about personality [2, 3], sentiments [4–6], and opinions [6].

Facebook (FB) has the current lead in social media market with approximately 1.86 bn monthly active users [1]. Apart from the marketing campaigns [7], publicity of movies [8], and current news updates [9], Facebook has been widely used by government bodies, police departments for community outreach, and safety concerns [10]. It is important to harness this powerful tool that supports quick two way interactions between the authority and the public quickly, establishing a mutual beneficial net. The interactions have been proved very effective at the times of emergencies [11, 12], and political elections as well [13].

M. A. Hossain (✉) · R. Daga · S. Chakrabarti
Institute of Engineering & Management, Kolkata, India
e-mail: a2hossain2000@gmail.com

R. Daga
e-mail: rashikadaga@gmail.com

S. Chakrabarti
e-mail: satyajit.chakrabarti@iemcal.com

S. Goswami
University of Calcutta, Kolkata, India
e-mail: saptarsi.goswami@iemcal.com

© Springer Nature Singapore Pte Ltd. 2019
J. K. Mandal et al. (eds.), *Contemporary Advances in Innovative and Applicable Information Technology*, Advances in Intelligent Systems and Computing 812,
https://doi.org/10.1007/978-981-13-1540-4_22

In our research work, data was mined from the official Facebook page maintained by Police Department of Kolkata, Kolkata Traffic Police (KTP), which is mainly a forum for Traffic Alerts and Public Grievances. With approximately 135,000 followers, the officials have been quite active in the last 2 years to increase the public engagement, effective problem solving via complaints on Facebook. Our research work aims to convert the unstructured data (data from Facebook posts) to structured data for better understanding, visualization of data and summarize the results in an effective manner. We also intend to highlight the views of the public via the reviews made on the KTP. This will help the viewers to get the crunch of the matter quickly, and more importantly, this will be of a great help to the Kolkata Police Department and other concerned authorities in analyzing the information that was being generated directly or indirectly via the page. It will help in increasing the quality of their content, and this brief overview will help in highlighting the shortcomings and positive results from the page. Besides the work of extracting information from preprocessed data after applying various techniques, there were other challenges that needed to be overcome to produce approximately accurate results.

2 Related Works

Social Networking websites have played a crucial role at the time of crisis [14, 15], helped improve efficiency of the organization, enhance the communication between government and citizens, and it increases transparency, and credibility in government [16–18]. Bertot et al. [19] in their study reveal that the social network websites like Facebook, Twitter puts more emphasis on the much-needed digital interaction between the government and the common mass, which in turn increase the engagement. It has been depicted how the posts on Facebook, shares, likes and comments, content sharing on government fan pages indirectly and directly helping the government know about the engagement and interests of the public [16].

Graham et al. have examined the use of social media tools by local governments and they conclude that 70% of the government bodies use social media tools and Facebook and Twitter are the most commonly used platforms, and they also discuss about the fact that most of the tweets and posts talk about special events and activities [20]. Sara Hofmann and co-authors analyze the Facebook sites of German local governments using a multi-method approach and depict the success of communication between the government and users in terms of frequency and polarity of citizen's reactions (sentiment analysis). Social Media offers the benefit of attaching multimedia features to a normal post to make it more attractive and gain popularity [21]. They conclude that posts that contain photos or videos are more liked or commented than those containing textual updates or a link. This shows the multimedia feature of Social media helps to capture the attention better.

Mossberger et al. [22] have studied the use of various interactive social network tools used in US and the work also suggests that "push" strategies are more

Table 1 Related works

Reference	Year	Objective
[24]	2016	User engagement, topic extraction, and classification from 4 Police departments (USA) Facebook pages and applying statistical methods to compare the posts and examine user and police interactions
[25]	2016	Analytic generalization to identify categories in government–citizen communication Karlstad municipality's FB page
[14]	2014	Content analysis of posts on The City of Calgary's Facebook page during Alberta flood to extract the main themes, checking comments on user postings to highlight significantly different themes
[26]	2015	Measure the impact of media and content types on stakeholders' engagement on Western European governments' FB pages
[27]	2015	Perform iterative content and web analysis on 31 'Information World Cities' websites and social media services and compare different social media services and types
[28]	2013	Examines the presence, usage and effectiveness of Egyptian government social media websites

dominant than networking and pull strategies in Facebook and Twitter. Magnusson's work [23] investigated the citizen–government interaction during flood by conducting analysis on two Facebook pages maintained by Local Government. The work also lists the type of posts and topics that catch user' attention and there is notable pattern that people tend to like and buzz about posts that are related to daily operations. Table 1 lists a few related works that have been carried out in this domain, along with their objective.

3 Proposed Methodology

This section lists the steps followed from collecting the data to the findings in a sequential manner.

Data Collection: Data from the Kolkata Traffic Police's Official Page on Facebook has been collected. The extracted data includes Page owner posts (posts made by KTP). Visitors posts, reviews by users, and comments on Page owner posts. Page Owner Posts were mined for a period of November 2016 to October 2017 for analysis purpose, visitor posts and reviews for the period of January 2017 to November 2017 were taken into consideration while developing algorithms.

Data Preprocessing: Posts that were posted in vernacular languages like Hindi or Bengali had to be removed manually from the database. Standard techniques for whitespace and punctuation removal, removal of stop words, link removal, and email id removal were applied on the messages posted by the users and KTP on the Facebook page. All the types of text cleaning, along with stemming was performed for the reviews.

User Engagement Statistics: Statistics such as like, comment and share count, average number of posts, posts on weekdays, increase in followers were determined to see if users are actually getting engaged from KTP's content.

Post Categorization: The types of posts are categorized into 4 types—Photo, Status, Video and Link. The status updates have been further classified into Traffic Updates and other updates. The traffic updates are further divided on the basis of reasons due to which the congestion takes place.

Complaint Count: Visitors' posts have been analyzed mostly to get quick information about the trend in the number of complaints posted by the visitors on a major issue that KTP is levying wrong cases on vehicles.

Sentiment Analysis: Sentiment Analysis was performed on the reviews extracted on a bimester basis. The "sentiment" package was used for classifying the polarity of the reviews as well as classifying emotions into six categories namely anger, joy, disgust, fear, sadness, and surprise.

Visualization: Visualization of congested places in Kolkata over a month has been depicted using Google maps for quick reference for users, and simple graphs and Word Cloud have been used for depicting the results obtained.

4 Method Overview

Data was extracted from the official Facebook page, KTP using the package Rfacebook developed in R (Version 3.3.2). It provides an interface to the Facebook API [29]. The data was a combination of unstructured (status, comments) as well as structured data (like count, comment count and share count). The reviews were extracted manually in our study. Attributes of page owner posts and visitors' posts that were used in developing the further algorithms are id of the post, textual status, like, comment and share count, and type of the post.

Removal of whitespaces, redundant words and punctuation errors was removed using R package "tm". Page followers on start and end of every month were obtained from getInsights function in Rfacebook package [29].

The textual updates from Page Owner posts were further divided into two categories—traffic updates and other updates not related directly to the primary concern of KTP (like Diwali Posts, Durga Puja Posts). Traffic updates were filtered on the basis of keywords like "Traffic", "update", "alert" and the top congested places of a month were further stored in another csv file, along with a count of their recurrence. All the traffic update posts were firstly checked whether they contain words like "Road", "Street", "flyover", etc., and two and three word strings were generated and checked if the location existed. Else, the strings were compared with a database of locations of Kolkata. The locations extracted were mapped on a Google Map. Traffic update posts were then classified on the basis of reasons of congestion, if mentioned in the post. This required typical text mining approach, along with a combination of pattern matching. Sentiment Analysis techniques are used to understand the feelings, and views of the users towards KTP, its initiatives and work. Based on the opinion words available in the "sentiment" package, a score was assigned to each feature. The package was used to determine the polarity and emotion of the reviews as mentioned in Sect. 3.

Complaints from visitors post were filtered similarly on the basis of keywords like "case", "prosecuted", and the relative percentage of complaints was noted for every month. Two kinds of complaints were considered in the analysis. First, complaints for wrong cases made by KTP for violating the traffic rules, and second, complaints made by users regarding other vehicles, that do not comply the rules or refuse to operate or provide service. R Packages "ggmap2", "WordCloud" has been used for visualization purpose.

5 Results

5.1 User Engagement Statistics

One of the primary objectives of the work is to deduce the citizen outreach and interaction of the common public with the local departments via social networking sites. The number of views a post gets, and the relative increase of the views would have accurately measured the citizen outreach. However, Facebook API does not allow the users to extract the view count of a post. Therefore, like, comment and share count are considered as surrogated indicators.

Frequency of posts

Table 2 shows the no. of posts, posts on weekend and average posts per day made by KTP from November 2016 to October 2017. This table gives a clear indication about the fluctuating frequency in posts by KTP. Along with irregular posting nature, it can be deduced that very few posts are made during the weekend relative to the total number of posts made in that particular month, hence lesser activity and interaction is observed during weekends compared to weekdays.

Like, Comment and Share Count

Table 3 shows the distribution of like count, comment count and share count per month. Highest number of likes, comments, and shares is recorded for the month of October, when the number of posts is 103. However, better engagement statistics are observed for the month of June, July and August owing to a large number of comments, like and share count out of lesser number of posts. Thus, we define a parameter—Interaction using likes as indicator for measuring the engagement is given by

$$\text{Interaction} = \frac{\text{Total Like Count of a particular month}}{\text{Total Page owner posts of a particular month}}$$

Table 2 No. of page owner posts, posts on weekend and average posts per day arranged month-wise

	Nov'16	Dec'16	Jan'17	Feb'17	Mar'17	Apr'17	May'17	Jun'17	Jul'17	Aug'17	Sep'17	Oct'17
Posts	83	60	70	55	72	54	54	59	88	36	39	103
Posts on weekend	25	9	6	10	3	5	1	3	3	5	7	6
Avg. posts/day	2.76	1.93	2.25	1.96	2.32	1.8	1.74	1.97	2.84	1.16	1.3	3.32

Table 3 Likes, comment and share count from Nov'16 to Oct'17

	Nov'16	Dec'16	Jan'17	Feb'17	Mar'17	Apr'17	May'17	Jun'17	Jul'17	Aug'17	Sep'17	Oct'17
Likes	1551	972	1190	1270	1000	1287	2279	12493	12669	16061	3674	19447
Comments	35	22	40	114	29	55	142	1061	988	491	179	2903
Shares	103	96	66	339	112	163	642	1899	1880	15435	1622	5094

Thus, highest interaction using likes as surrogated indicator is recorded for August 2017, whereas lowest interaction of the public with the official page is noted for December 2016. Similarly, we can obtain the interaction using comments and shares as surrogated indicators for measuring engagement.

Video updates also get a high amount of outreach, however, the frequency of video posts is the least. The multimedia features of social networking website are indeed successful in catching users' attention and expanding the online community. Based on the statistics obtained, KTP is focused on increasing the reach of its posts by increasing the number of updates, maintaining a ratio between status and photo updates, and attempt to post videos in a month to create a much-needed buzz. Figure 1 highlights the percentage breakup of likes obtained from various types of posts and videos and photos are more popular than the other two types. Statuses, which are mainly composed of traffic updates, do not generate a high number of likes, comments or shares.

"No Engagement" Posts

Figure 2 shows the percentage of post updates by KTP that do not generate a substantial user engagement. Any post that had likes less than 10 was taken in the category of generating no substantial user engagement. Moreover, months with higher percentage of posts that do not cross threshold have lesser likes, comments and shares compared to other months. February, March, April, May and September have not gained significant user engagement as depicted by Fig. 2. In August, KTP posted a video that got viral, due to which the users seemed interested in KTP's

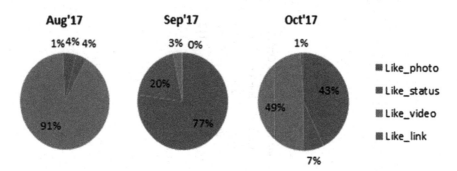

Fig. 1 Percentage of likes on different kinds of posts for the months of August, September and October, 2017

Fig. 2 Percentage of posts that do not generate significant user engagement

posts, and it saw lesser percentage of "No Engagement" posts, and high increase in number of followers.

Trend in the Increase of Followers

The number of followers in November 2016 was 95,839, and had a count of 134,355 at the end of October. This increase and the trend in increase give knowledge about the reach and popularity of the page. Figure 3 shows the increase in number of followers every month. The increase pattern definitely shows the fact that the engagement has increased till August, and September, and October show lesser engagement.

5.2 Complains from Visitors' Post

In our study, we have extracted two types of complaints as mentioned in Sect. 4. Type 1 refers to complaints made for KTP imposing wrong cases on individuals for violation of traffic rules. Type 2 refers to complaint made on specific vehicles for violation of rules, bad behavior, and denial of service. Figure 4 shows the percentage of Type 1 and Type 2 complaints in visitors' posts. The graph of Type 1 decreases from January to October 2017, showing that KTP is making an effort to reduce the mistake of lodging wrong complaints.

5.3 Traffic Updates

KTP makes updates on the current Traffic situation in Kolkata. In our study, we extracted the number of posts that concentrate on traffic. The number of status

Fig. 3 Increase in the number of followers per month

Fig. 4 Percentage of Type 1 and Type 2 complaints

updates have increased from June to October (depicted by Table 3), indicating that KTP is focusing more on the primary objective of informing people about the congestion as fast as possible through this media tool. From the congested places, certain locations like Chingrighata, R.R. Avenue occurred in the top congested locations' list in almost all months. The list is quite useful for finding out the trend in traffic control of locations and if KTP is paying attention to frequently congested places. In our work, a google map is also produced that shows the top congested places in a month. Figure 5 shows the congested places in the month of October, 2017. Places with a bigger spot were more frequently jammed than the ones with a smaller mark. In our study, we are able to find out the reason for congestion, if mentioned by KTP.

Fig. 5 Google map representing the congested places of Kolkata in Oct'175.4. Sentiment Analysis

The reasons for congestion are religious, cultural and political procession, dispersal of school, and construction work, where political gatherings and dispersal of School are the most frequent reasons for Congestion.

We have applied sentiment analysis on user reviews of the Kolkata Traffic Police using Emotion Categorization, Polarity Categorization, and Word Cloud as mentioned in Sect. 3. The algorithm was applied on a bimester for better results. The user reviews frequency shows a progressive trend with the frequency of reviews increasing, culminating during the period of September–October and then sharply falling back for the next bimester. Positive sentiment for the Police force is highest during the time of September–October due to a number of festivals like Durga Puja, Diwali as seen in Fig. 6. The positive reviews are mainly on accounts of excellent traffic and crowd management as the streets are very busy during this time. Many reviews also empathize with the police force for their dedication as these holidays are usually categorized as a family event. Positive sentiments for other months can be generalized to factors like effective traffic control, manual clearing up of waterlogged areas, empathy for police force during extreme weather conditions and individual experiences like car breakdown or lost items (Figs. 7 and 8).

Figures 9 and 10 delineate the polarity of sentiment expressed in the reviews. The months of March and April show a low sentiment score mainly due to harassment, bad traffic management, corruption and false cases.

Cases of harassment include unprofessional behavior such as public humiliation and the use of profane language. False cases include wrong traffic violation cases imposed on individuals. Complaints about bad traffic management deals with violations committed by buses, auto-rickshaws as well as space encroachment by hawkers and peddlers leading to congested roads. Corruption reviews talk about the bribery and extortions made by some police officers.

The word cloud displayed in Fig. 8 is used for visually portraying keywords or tags more frequently used by the users. The size of the words in the word cloud represents the frequency of words. Reviews on bikes are greater than that of taxis, followed by autos. The word "good" is most often characterized with the KTP. However it should also be noted that words like "money" and "fine" are also used which indicates bribery and fines charged due to violation of traffic rules. "Drivers" is a common word when it comes to roads and traffic so it occupies a large space; users most often use it to describe the unprofessional and rash driving of bus and auto-rickshaw drivers. "Drivers" is also used in the case of absenting Uber or Ola drivers frequently.

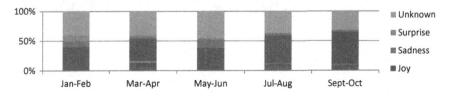

Fig. 6 Emotion categorization from January to October, 2017

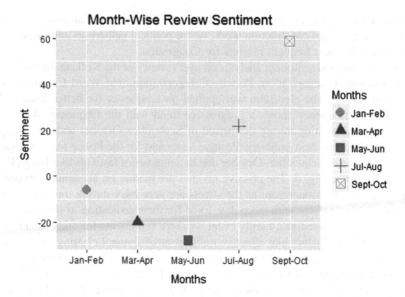

Fig. 7 Sentiment score on reviews from Jan'17–Oct'17 on a bimester basis

Fig. 8 Word cloud on
reviews generated from
Jan'17–Nov'17

Fig. 9 Polarity graph for
reviews of Mar'17–Apr'17

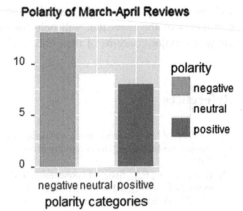

Fig. 10 Polarity graph for
reviews of Jul'17–Aug'17

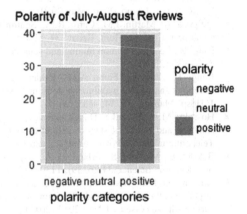

6 Conclusion

The local police department plays an important role in dispersing information about
the city, and this role has only become bigger with the advent of social media. In
this paper, we have analyzed the interaction of Kolkata Traffic Police and the users
on the Facebook platform and have proposed a methodology to visually represent
large amounts of information that will simplify the understanding of both the
concerned authorities as well as the civilians. Our proposed method, first analyzes
the user interaction, which will help the concerned authorities to effectively utilize it
to increase interaction and awareness. It has been established that the type of posts
—like photo and video updates is enhancing the user engagement, and the multimedia features of Facebook are attracting audience. Secondly, it tracks the complaints from users, which will help the police to serve better; thirdly it plots the
traffic updates that are helpful for the public in knowing the congested places,

without scrolling through the textual updates all the time. Lastly, it analyzes the sentiments of the users that will help the police force to rectify their flaws, and know the major concerns of the public.

References

1. Number of social media users worldwide from 2010 to 2021 (in billions). https://www. statista.com/statistics/278414/number-of-worldwide-social-network-users. Accessed 15 Nov 2017
2. Ross, C., Orr, E.S., Sisic, M., Arseneault, J.M., Simmering, M.G., Orr, R.R.: Personality and motivations associated with Facebook use. Comput. Hum. Behav. **25**(2), 578–586 (2009)
3. Amichai-Hamburger, Y., Vinitzky, G.: Social network use and personality. Comput. Hum. Behav. **26**(6), 1289–1295 (2010)
4. Pak, A., Paroubek, P.: Twitter as a corpus for sentiment analysis and opinion mining. In LREc, vol. 10, no. 2010 (May 2010)
5. Ortigosa, A., Martín, J.M., Carro, R.M.: Sentiment analysis in facebook and its application to e-learning. Comput. Hum. Behav. **31**, 527–541 (2014)
6. Pang, B., Lee, L.: Opinion mining and sentiment analysis. Found. Trends. Inf. Retrieval **2**(1–2), 1–135 (2008)
7. Kang, J., Tang, L., Fiore, A.M.: Enhancing consumer–brand relationships on restaurant Facebook fan pages: maximizing consumer benefits and increasing active participation. Int. J. Hosp. Manage. **36**, 145–155 (2014)
8. Hossain, M.A., Rashika, D., Goswami, S.: Integration of classical features, social media response and power of stars to predict the box office sales. In International Conference on Telecommunication, Power Analysis, and Computing Techniques (April 2017)
9. Bakshy, E., Messing, S., Adamic, L.A.: Exposure to ideologically diverse news and opinion on Facebook. Science **348**(6239), 1130–1132 (2015)
10. Kim, K., Oglesby-Neal, A., Mohr, E.: 2016 Law enforcement use of social media survey. http://www.theiacp.org/Portals/0/documents/pdfs/2016-law-enforcement-use-of-social-media-survey.pdf. Accessed 24 Nov 2017 (2017)
11. Muralidharan, S., Rasmussen, L., Patterson, D., Shin, J.H.: Hope for Haiti: an analysis of Facebook and Twitter usage during the earthquake relief efforts. Public Relat. Rev. **37**(2), 175–177 (2011)
12. Bird, D., Ling, M., Haynes, K.: Flooding Facebook-the use of social media during the Queensland and Victorian floods. Aust. J. Emerg. Manage. **27**(1), 27 (2012)
13. Vitak, J., Zube, P., Smock, A., Carr, C.T., Ellison, N., Lampe, C.: It's complicated: Facebook users' political participation in the 2008 election. CyberPsychol. Behav. Soc. Network. **14**(3), 107–114 (2011)
14. Magnusson, M.: Information seeking and sharing during a flood-a content analysis of a local government's facebook page. In Proceedings of the European Conference on Social Media. Academic Conferences Limited, Reading, UK, pp. 305–311 (July 2014)
15. Muralidharan, S., Rasmussen, L., Patterson, D., Shin, J.H.: Hope for Haiti: an analysis of Facebook and Twitter usage during the earthquake relief efforts. Public Relat. Rev. **37**(2), 175–177 (2011)
16. Mergel, I.: A framework for interpreting social media interactions in the public sector. Gov. Inform. Q. **30**(4), 327–334 (2013)
17. Chang, A.M., Kannan, P.K.: Leveraging Web 2.0 in Government. IBM Center for the Business of Government, Washington, DC (2008)

18. Picazo-Vela, S., Gutiérrez-Martínez, I., Luna-Reyes, L.F.: Understanding risks, benefits, and strategic alternatives of social media applications in the public sector. Gov. Inform. Q. **29**(4), 504–511 (2012)

19. Bertot, J.C., Jaeger, P.T., Grimes, J.M.: Using ICTs to create a culture of transparency: e-government and social media as openness and anti-corruption tools for societies. Gov. Inform. Q. **27**(3), 264–271 (2010)

20. Graham, M., Avery, E.J.: Government public relations and social media: an analysis of the perceptions and trends of social media use at the local government level. Public Relat. J. **7**(4), 1–21 (2013)

21. Hofmann, S., Beverungen, D., Räckers, M., Becker, J.: What makes local governments' online communications successful? Insights from a multi-method analysis of Facebook. Gov. Inform. Q. **30**(4), 387–396 (2013)

22. Mossberger, K., Wu, Y., Crawford, J.: Connecting citizens and local governments? Social media and interactivity in major US cities. Gov. Inform. Q. **30**(4), 351–358 (2013)

23. Magnusson, M.: Facebook usage during a flood—a content analysis of two local governments'. Facebook pages (2016)

24. Huang, Y., Huo, S., Yao, Y., Chao, N., Wang, Y., Grygiel, J., Sawyer, S.: Municipal police departments on Facebook: what are they posting and are people engaging? In Proceedings of the 17th International Digital Government Research Conference on Digital Government Research, pp. 366–374. ACM (June 2016)

25. Bellström, P., Bellström, P., Magnusson, M., Magnusson, M., Pettersson, J.S., Pettersson, J. S., Thorén, C., Thorén, C.: Facebook usage in a local government: a content analysis of page owner posts and user posts. Transform. Gov: People Process. Policy **10**(4), 548–567 (2016)

26. Bonsón, E., Royo, S., Ratkai, M.: Citizens' engagement on local governments' Facebook sites. An empirical analysis: the impact of different media and content types in Western Europe. Gov. Inform. Q. **32**(1), 52–62 (2015)

27. Mainka, A., Hartmann, S., Stock, W.G., Peters, I.: Looking for friends and followers: a global investigation of governmental social media use. Transform. Gov. People Process. Policy **9**(2), 237–254 (2015)

28. Abdelsalam, H.M., Reddick, C.G., Gamal, S., Al-Shaar, A.: Social media in Egyptian government websites: presence, usage, and effectiveness. Gov. Inform. Q. **30**(4), 406–416 (2013)

29. Barbera, P., Piccirilli, M., Geisler, A., van Atteveldt, W.: Rfacebook: access to Facebook API via R. R package version 0.6.15. https://CRAN.R-project.org/package=Rfacebook (2017)

Sentiment Analysis Based Potential Customer Base Identification in Social Media

Sanjay Goswami, Satrajit Nandi and Sucheta Chatterjee

1 Introduction

In today's world, most common and popular platform for advertisement is the internet because of its affordability and timeliness. Advertisement in intangible formats can be spread much more quickly rather than tangible one. This is one of the aspects of online advertisements where advertiser can quickly spread advertisements among the globe [1]. As being intangible the spreading cost is much lesser rather than tangible.

In these days, people are getting tech-savvy, there are a huge number of people are using internet. And among them a huge number of people are attached to social networking platforms like Facebook, twitter, etc. Hence if the advertiser target internet as a platform for advertisements so he or she can easily get a huge number of audience.

But there is a negative aspect of getting so many crowds. Getting so many crowds means getting a huge number of viewers with respect to a particular advertisement though there every viewer may not be a potential customer. Here the advertiser has to keep one thing in his or her mind, i.e., "Cost per Action" (CPA). This is also known as "Cost per Acquisition" [2]. This is the most common online advertising model where the pricing depends on a specific acquisition. As an example if an internet user click on an advertisement, then the advertiser has to pay

S. Goswami (✉)
Narula Institute of Technology, Kolkata, India
e-mail: sanjaygoswamee@gmail.com

S. Nandi
Ideal Institute of Engineering, Kalyani, India
e-mail: satrajit.nandi@gmail.com

S. Chatterjee
National Institute of Technical Teacher's Training and Research, Kolkata, India
e-mail: sucheta.chatterjee25@gmail.com

© Springer Nature Singapore Pte Ltd. 2019
J. K. Mandal et al. (eds.), *Contemporary Advances in Innovative and Applicable Information Technology*, Advances in Intelligent Systems and Computing 812, https://doi.org/10.1007/978-981-13-1540-4_23

to that advertisement promoting platform. This model can be further decomposed in specific pricing models based on its action type. Like "Cost per Click" (charged for clicking on advertisement), "Cost per Impressions" (charged for viewing an advertisement), "Cost per Lead" (charged for every sign-up) etc. These days majority of advertisement platform and tools providers like Google, Facebook, Bing follows these online advertising pricing models.

So the advertiser must identify the proper target audience for their advertisement else the investment could be wasted. Every online advertiser has to find the *affinity* value for every individual advertisement. *Affinity* is a natural linking between something. It means a relationship between one entity with another entity because of their common interests. It can be described as a quality which makes things or entity suited to each other. This *affinity* identification is a very hard job which has higher chance of improvements. This is one of the major issues where advertiser loses a lot of money due to miss-identification of *affinity* values.

Social network's also become a very popular advertisement platform because of its huge user data statics. The advertiser can access those data to find the suitable zone to publish advertisement. But the main problem is the volume of these data itself, which is very hard to be analyzed by human. Because when advertiser starts analysis they have to form and test the data, at the same time they have to map the data with every test action. It means whenever they change their marketing tactics, they have to recalculate and compare the data with previous data. Also they have to track their competitor's action. This will also help to change the marketing tactics more accurately. These days there are many tools available which helps to analysis these market data. But all these tools work in a same way. They create quick reports according to the benchmarks, defined by advertiser. Some of the tools [3] also help to reuse posts. But all of them do a common job that is measuring total engagement of audience. Some of the popular tools are: Google analytics (measures traffic to the website), Keyhole (measure social networks impact), Buffer (measure engaged people in social network post), Quintly (compare profile with competitors) etc.

There are more things which add some portion of cost in online advertisement; like content development and regularity for content promotion. Online advertisement requires digital assets like text, image, audio, video, etc., files. But it is not necessary to develop each content from scratch. There are plenty royalty-free contents also known as Creative Commons which can be used to develop an advertisement. This reuse process can make the development process faster and can cut the costs of development. Now there is one human behavioral factor which advertiser has to keep in minds, that is regularity in activity; by virtue of that advertiser can keep in touch with the audience. As an example, if an advertiser push content very often to the promoting webpage, the viewer of that page will become lesser rather than the page which pushes more content very frequently. It happens because one of the human behavior, attachment. But pushing informative content more frequently requires more engagement which will add some extra cost for promotion.

Today's online advertisement mechanism can guide advertiser in such a way where they can promote to reach their targeted audience. This system mainly depends on the user data which gathered from the web. But this is a big drawback of present system. Whenever anyone uses past data or existent raw data to generate a statistical report which can be used for identification; they can only generate reports for those things which are laid between previous data elements. That means if an advertiser wants to promote an item which is for the first time to be launched in the market, advertiser cannot get any statistical data for promotion.

2 Literature Review

Authors Patil and Atique [4] described the importance of sentiment analysis and option mining. Here the authors investigate about the challenges of sentiment analysis. They also provide few case studies where this approach is used. The main goal of their research is to give an overview of sentiment analysis in present day. As conclusion they also found that this is mainly based on machine learning approach.

Author Corley et al. [5] described that in these days Internet marketing is a rapidly growing research area. They survey a lot and analyzed 411 articles published from the year 1994. They described those methodologies which have been published along with some methodologies which needs further exploration. Author also outlined few research topics like, "business models of online marketing", "the future of search strategies", "the internet advertising landscape", "evaluation of online performance", etc. There entire work motivates researcher for further exploration in this research domain.

Author Zeng et al. proposed a methodology [6] in the patent where they can generate a ranking among the reputation corresponding to the advertised product or service or seller's reputation. This ranking value generates through a mathematical function. This reputation also depends with the feedback of customers. This proposed methodology basically helps websites to gain profits from advertisements. Websites that contains advertisement can easily follow this method to rank among those advertisements which are already present on their website to generate more profits.

Author Xing et al. described the importance of search engine optimization [7] in advertising. Their research aims to bring impact on advertising market through search engine optimization by their analytical model. The goal of this research work is to help and informed managers to make decisions in online advertising.

Author Liu and Zhang proposed an automatic opinion mining and summarization system [8] which will summarize the public opinion without biasness. Human analyzed information text is generated with their own biases. Authors describe a way to find key elements in opinions. Also they discuss about the issue of spam or fake reviews and their detection.

Authors Crespo et al. suggested a framework [9] which can classify the interaction between the customer in feedback section. These days' people use web for their communication media and for knowledge acquisition. Hence it is very important in customer relationship model that what the customer think about.

Authors have done data mining and natural language processing for classification and analysis.

Authors Liu et al. proposed a system and developed a prototype [10] which can observe the opinion from consumers. There framework can analyze and compare between each consumer's feedback of competing products. And their developed system can clarify the strength and weakness of each product on the view of consumers. This system helps both, consumer and manufacturer. Consumer can compare between same types of products. In other side, manufacturers also gather the market information for a particular product and they can upgrade it for rectification, based on pros and cons in customer review section.

Dr. Gandhi proposes a research framework [11] based on multi criteria decision making method. Their main goal is to identify the potential influencers from social media. These days' social media users become major online contributors. So identification of most influencer is very necessary to business enterprises so that they could take necessary action to engage and maintain that potential influencer.

3   Objective

The main objective of this research work is to identify and create a cluster of groups of similar kind of interests. This will also identify group of consumer according to a particular product category. After customer clustering it will identify those consumers who are already satisfied or unsatisfied with equivalent product by judging their negative or positive biases in their posts. This will help advertiser to identify where they can advertise to gain maximum profit.

4   Methodology

The proposed research work consists of four sequenced phases. These are described below:

Phase 1: Clustering of posts related to similar products. (Product type or service type based clustering)

Phase 2: Identification of each type of products in each cluster.

Phase 3: Sentiment identification of users related to a particular product. (Negative or positive biases identification. i.e. Sentiment wise sub-clustering within clustering group)

Phase 4: Sentiment based hierarchical clustering within interest group.

Additional Phase: Automation tool development for automating promotion of advertisements.

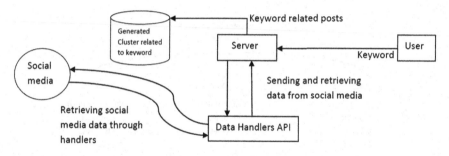

Fig. 1 Keyword-related cluster generation

In the first phase we have to fetch the entire similar type product against advertiser's own product related posts from the social media. Here we have to use YouTube, Twitter and Facebook's data handler API to access public's posts from social network (Fig. 1). After gathering all of those posts we have to identify similar type of category for the targeted category. Here we have to use semantic search technique to find the related product list. As an example if the targeted product is a skirt, then at first we have to use semantic search to find which category it belongs to, then we have to create a cluster of that category (Fig. 2). In this case, this target is in the category of clothing section. So we have to gather all the posts related to clothing. And make a cluster of clothing related posts.

In second phase we have to identify each type of product in the cluster which is formed in phase number one. For this purpose we will use NoSql, because these data is purely unstructured data. Here we have to identify all of the variations. Basically in this phase sub cluster will be generated from parent cluster. As an example if the parent cluster is clothing section, then we have to identify each type of product category in it. Like men's t-shirt, woman's t-shirt, men's pant, woman's pant, women skirt etc.

In the third phase, we have to read all the post related to targeted product according to biasness of public. That means we have to use text classifier for

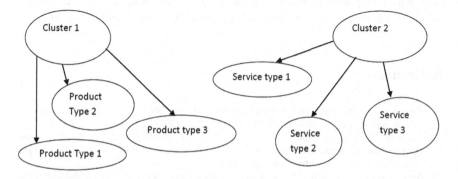

Fig. 2 Identification of products in each cluster

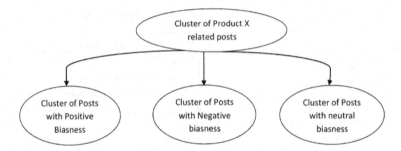

Fig. 3 Sentiment-based sub-clustering

emotion identification in posts. In present days there is a classifier, MaxEnt [12] which is used to classify text and speech in natural language processing. This is also known as Maximum Entropy Text Classifier. Commonly there are three types of sentiment which we will consider; these are positive, negative and neutral.

In fourth phase, we have to create sub cluster according to sentiment for each unique cluster generated in second phase (Fig. 3). So in this cluster we can divide every unique cluster into three different clusters according to their positive, negative and neutral biases in their posts. As an example, if our targeted product is skirt, then we can get the cluster, one who likes skirts, one who dislike skirts and one who do not dislike it but at same time do not like it also.

At last, in prototype section we have to build a prototype where these phases are implemented and advertiser can automate advertisement is social media.

5   Conclusions

This proposed system is expected to identify those clusters or locations of consumers who have positive biases towards the target product or service. But consumer, who have positive biases, may not be the actual buyers. Hence it is required to identify a hierarchical cluster of products or services based on consumer biases so that actual consumer base can be identified who can be potential buyers.

References

1. Lua, A.: 7 social media analytics and reporting tips for becoming a data-savvy marketer. https://blog.bufferapp.com/learn-social-media-analytics (2017)
2. Spooner, J.: Why cost per acquisition is the only metric that really matters—social media explorer. https://socialmediaexplorer.com/content-sections/tools-and-tips/why-cost-per-acquisition-is-the-only-metric-that-really-matters/
3. Meyer, A.: A list of the top 25 social media analytics tools. http://keyhole.co/blog/list-of-the-top-25-social-media-analytics-tools/ (2016)

4. Patil, H., Atique, M.: Sentiment analysis for social media: a survey. In: Information Science and Security (ICISS), IEEE, Seoul, South Korea, pp. 389–392 (2015)
5. Corley, J., Jourdan, Z., Ingram, W.: Internet marketing: a content analysis of the research. Electron. Mark. **23**, 177–204 (2013)
6. Zeng, H., Lin, C., Han, D., Zhang, B., Chen, Z., Wang, J.: Ranking online advertisement using product and seller reputation (2008)
7. Xing, B., Lin, Z.: The impact of search engine optimization on online advertising market. In: 8th International Conference on Electronic Commerce: The New E-Commerce—Innovations for Conquering Current Barriers, Obstacles and Limitations to Conducting Successful Business on the Internet, ACM, New Brunswick, Canada, pp. 1–11 (2006)
8. Liu, B., Zhang, L.: A survey of opinion mining and sentiment analysis. In: Aggarwal, C., Zhai, C. (eds.) Mining Text Data, pp. 415–463. Springer-Verlag, New York (2012)
9. García-Crespo, Á., Colomo-Palacios, R., Gómez-Berbís, J., Ruiz-Mezcua, B.: SEMO: a framework for customer social networks analysis based on semantics. J. Inform. Technol. **25**, 178–188 (2010)
10. Liu, B., Hu, M., Cheng, J.: Opinion observer: analyzing and comparing opinions on the web. In: 14th International Conference on World Wide Web, ACM, Chiba, Japan, pp. 342–351 (2005)
11. Gandhi, M., Muruganantham, A.: Potential influencers identification using multi-criteria decision making (MCDM) methods. In: 3rd International Conference on Recent Trends in Computing (ICRTC), Elsevier, pp. 1179–1188 (2015)
12. Vryniotis, V.: Machine learning tutorial: the max entropy text classifier. Datumbox. http://blog.datumbox.com/machine-learning-tutorial-the-max-entropy-text-classifier/

Author Index

© Springer Nature Singapore Pte Ltd. 2019
J. K. Mandal et al. (eds.), *Contemporary Advances in Innovative and Applicable Information Technology*, Advances in Intelligent Systems and Computing 812,
https://doi.org/10.1007/978-981-13-1540-4

Printed in the United States
By Bookmasters